Workbook for Auto Collision Repair and Refinishing

Second Edition

by

Michael Crandell

Publisher
The Goodheart-Willcox Company, Inc.
Tinley Park, IL
www.g-w.com

Copyright © 2017
by
The Goodheart-Willcox Company, Inc.

All rights reserved. No part of this work may be reproduced, stored, or transmitted in any form or by any electronic or mechanical means, including information storage and retrieval systems, without the prior written permission of The Goodheart-Willcox Company, Inc.

Manufactured in the United States of America.

ISBN 978-1-63126-401-6

3 4 5 6 7 8 9 – 17 – 21 20 19 18 17

The Goodheart-Willcox Company, Inc. Brand Disclaimer: Brand names, company names, and illustrations for products and services included in this text are provided for educational purposes only and do not represent or imply endorsement or recommendation by the author or the publisher.

The Goodheart-Willcox Company, Inc. Safety Notice: The reader is expressly advised to carefully read, understand, and apply all safety precautions and warnings described in this book or that might also be indicated in undertaking the activities and exercises described herein to minimize risk of personal injury or injury to others. Common sense and good judgment should also be exercised and applied to help avoid all potential hazards. The reader should always refer to the appropriate manufacturer's technical information, directions, and recommendations; then proceed with care to follow specific equipment operating instructions. The reader should understand these notices and cautions are not exhaustive.

The publisher makes no warranty or representation whatsoever, either expressed or implied, including but not limited to equipment, procedures, and applications described or referred to herein, their quality, performance, merchantability, or fitness for a particular purpose. The publisher assumes no responsibility for any changes, errors, or omissions in this book. The publisher specifically disclaims any liability whatsoever, including any direct, indirect, incidental, consequential, special, or exemplary damages resulting, in whole or in part, from the reader's use or reliance upon the information, instructions, procedures, warnings, cautions, applications, or other matter contained in this book. The publisher assumes no responsibility for the activities of the reader.

The Goodheart-Willcox Company, Inc. Internet Disclaimer: The Internet resources and listings in this Goodheart-Willcox Publisher product are provided solely as a convenience to you. These resources and listings were reviewed at the time of publication to provide you with accurate, safe, and appropriate information. Goodheart-Willcox Publisher has no control over the referenced websites and, due to the dynamic nature of the Internet, is not responsible or liable for the content, products, or performance of links to other websites or resources. Goodheart-Willcox Publisher makes no representation, either expressed or implied, regarding the content of these websites, and such references do not constitute an endorsement or recommendation of the information or content presented. It is your responsibility to take all protective measures to guard against inappropriate content, viruses, or other destructive elements.

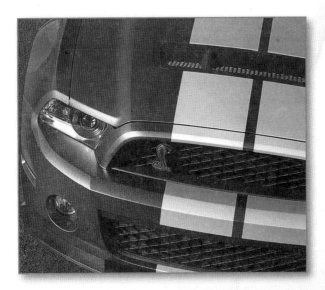

Introduction

This workbook is designed for use with the textbook *Auto Collision Repair and Refinishing*. As you complete the questions and problems in this workbook, you can review the facts and concepts presented in the textbook.

Each chapter of the workbook includes objectives and instructions. Several types of questions and problems are given in each chapter. The various types of questions include matching, identification, multiple choice, completion, and short answer. Also included in the workbook are a number of jobs that outline common collision repair and refinishing procedures.

The workbook chapters correspond to the chapters in the textbook. After studying a chapter in the textbook, complete the corresponding workbook questions carefully and accurately. Then, complete the workbook jobs that are applicable to the chapter you are studying. This will allow you to apply the information learned in the chapter in hands-on applications.

Auto Collision Repair and Refinishing is an indispensable resource for anyone preparing for a career in collision repair and refinishing, as well as experienced technicians preparing for the ASE Collision Repair and Refinish certification tests. Studying the *Auto Collision Repair and Refinishing* textbook and successfully completing this workbook will help you develop a solid background in collision repair and refinishing.

Contents

		Workbook Page	Textbook Page
Chapter 1	Introduction to Collision Repair	9	3
Chapter 2	Safety	15	17
Chapter 3	Vehicle Construction	21	41
Chapter 4	Fundamentals of Collision Damage	27	59
Chapter 5	General-Purpose Tools and Equipment, Service Information	33	81
Chapter 6	Fasteners	41	101
Chapter 7	Welding and Cutting	47	115
Chapter 8	Nonstructural Repair Tools, Equipment, and Materials	55	145
Chapter 9	Nonstructural Panel Repair	63	173
Chapter 10	Bolted Nonstructural Part Replacement	71	207
Chapter 11	Welded and Bonded Nonstructural Panel Replacement	77	237
Chapter 12	Plastic Repair	85	269
Chapter 13	Glass	91	295
Chapter 14	Unibody/Frame Straightening Equipment	97	317
Chapter 15	Measurements	103	341
Chapter 16	Unibody Straightening	109	369
Chapter 17	Full-Frame Repair	117	397
Chapter 18	Structural Component Replacement	123	427
Chapter 19	Steering and Suspension	129	457
Chapter 20	Electrical System	135	495
Chapter 21	Brakes	141	519
Chapter 22	Cooling, Heating, and Air Conditioning Systems	147	535
Chapter 23	Power Train	153	553
Chapter 24	Restraint Systems	159	575
Chapter 25	Refinishing Tools and Equipment	163	595
Chapter 26	Refinishing Materials	167	627
Chapter 27	Paint Mixing and Reducing	173	655
Chapter 28	Spray Technique	179	673
Chapter 29	Surface Preparation	185	697
Chapter 30	Color Matching	193	729
Chapter 31	Paint Application	197	749
Chapter 32	Specialty Painting	201	777
Chapter 33	Detailing	205	793
Chapter 34	Estimating	209	817
Chapter 35	ASE Certification	213	851
Chapter 36	Employment Strategies and Employability Skills	217	861

		Workbook Page
Job 1	Safety	221
Job 2	Vehicle Basics	227
Job 3	MIG Welding Basics	233
Job 4	MIG Welding Practice	237
Job 5	Metal Finishing Skills	241
Job 6	Removing and Installing Bolted Panels	245
Job 7	Panel Splicing	253
Job 8	Flexible Plastic Repair	255
Job 9	Measurements	259
Job 10	Removing and Installing Short Arm/Long Arm Suspension Components	265
Job 11	Removing and Installing a MacPherson Strut Assembly	271
Job 12	Removing and Installing a Radiator	275
Job 13	Removing and Installing a Drive Axle	277
Job 14	Electrical Practice	279
Job 15	Masking	285
Job 16	Mixing, Applying, and Sanding Body Filler	289
Job 17	Reducing Paint	293
Job 18	Spray Gun Basics	297
Job 19	Spray Gun Practice	301
Job 20	Spraying Basecoat/Clearcoat Paints	303
Job 21	Preparing for a Clearcoat Blend	309
Job 22	Buffing	313
Job 23	Applying Pinstripes	317

Instructions for Answering the Workbook Questions

Each chapter in this workbook directly correlates to the same chapter in the text. Before answering the questions in the workbook, study the assigned chapter in the text and answer the end-of-chapter review questions. Then, review the objectives at the beginning of each workbook chapter. Try to complete as many workbook questions as possible without referring to the textbook. Then, use the text to complete the remaining questions.

A variety of questions are used in the workbook including multiple choice, completion, identification, short answer, and matching. These questions should be answered in the following manner:

Multiple Choice

Select the best answer and write the correct answer in the blank.

_____B_____ 1. A permanent change in metal grain arrangement is called _____ deformation.
  A. elastic
  B. plastic
  C. internal
  D. unworking

Completion

In the blank provided, write the word or words that best complete the statement.

_____blocking_____ 2. During a tension pull, a technician should insert _____ under the suspension of the vehicle to keep the vehicle at the correct height and prevent suspension sag.

Identification

Identify the components indicated on the illustration or photograph accompanying the question.

3. Identify the parts of a spray gun shown in the following illustration.

   A. Air cap assembly
   B. Fluid tip
   C. Fluid needle
   D. Pattern adjustment valve knob
   E. Fluid control knob
   F. Air inlet
   G. Paint cup
   H. Vacuum tube
   I. Air vent
   J. Trigger
   K. Air nozzle

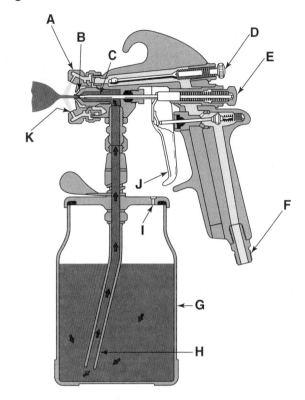

Short Answer

Provide complete responses to the statements.

4. Explain how to prevent *chasing the dent*.

   To prevent chasing the dent, tighten any floppy metal around the area to be filled by making a shrink fence. A shrink fence is simply a line of small kinks made with a sharp pick hammer. The shrink fence tightens the floppy metal in the low-crown area, allowing the filler to be sanded level.

Matching

Match the term in the left column with its description in the right column. Place the corresponding letter in the blank.

For questions 5–10, match the following vehicle paint problems with their correct description. Each answer will be used only once.

__D__ 5. Appears as reddish brown grains on bare steel.

__E__ 6. Gradual change in color due to paint pigment breakdown.

__B__ 7. Weather-beaten paint with no gloss due to paint binder breakdown.

__F__ 8. Creates a weak, flakey area with bubble-like paint blisters on the surface.

__C__ 9. Caused by loss of adhesion between the topcoat and primer.

__A__ 10. Begins as tiny slits in the paint surface often caused by UV exposure.

A. Cracking
B. Chalking
C. Peeling
D. Surface rust
E. Fading
F. Internal rust

Other Types of Questions

When other types of workbook questions are presented, follow the specific instructions that accompany the problem.

Instructions for Performing Workbook Jobs

The jobs in this workbook supplement the material in the textbook by outlining various hands-on activities. Before starting any job, read through the entire assignment and discuss the procedure with your instructor. It is also important to read the related chapters in the textbook and review all pertinent safety information.

The jobs in this workbook are numbered primarily for ease of reference. The numbers do not necessarily dictate the order in which the jobs must be performed. Some jobs can be done as part of other, more complex jobs. The order in which the jobs are performed is entirely up to your instructor.

Some jobs may take more than one class period to complete. When this occurs, be sure to inform your instructor so your project can be stored properly until you are able to resume work.

As you complete each step in a job, place a check mark in the corresponding box. This will help you keep track of your progress. If any of the steps in a given job do not apply to the particular vehicle or assembly you are working on, mark N/A, for not applicable, by the box. When you finish an entire job, ask your instructor to inspect your work and initial your completed job sheet.

Name _____ Date _____ Class _____

Chapter 1

Introduction to Collision Repair

Learning Objectives

After studying this chapter, you will be able to:
- Summarize the typical steps in the collision repair process.
- Compare the various types of body shop ownership.
- Describe the types of jobs available in the collision repair industry.
- Explain the educational opportunities available to those who are interested in a career in collision repair.

Carefully read Chapter 1 of the textbook and then answer the following questions in the space provided.

_____ 1. A vehicle that is damaged so badly that repair costs exceed the value of the vehicle is called a(n) _____.

_____ 2. *True or False?* An older vehicle, which is generally worth less than a newer vehicle, will be declared a total loss with less damage than its newer counterpart.

_____ 3. A vehicle's owner can buy _____ to pay for damages caused by the vehicle in the event of an accident.

_____ 4. Mr. Smith carries only liability insurance. If Mr. Smith hits Mrs. Wilson's vehicle, damage to Mr. Smith's vehicle is paid for by _____.
　A. the insurance company
　B. Mr. Smith
　C. Mrs. Wilson
　D. an auto body shop

_____ 5. In the example situation above, if Mr. Smith had _____ also, the insurance company would cover the repair cost of Mr. Smith's vehicle.

_____ 6. *True or False?* A deductible is the part of the repair cost that the vehicle owner is responsible for paying under comprehensive coverage; the insurance company pays the remaining repair cost.

_____ 7. A(n) _____ is made when an insured person suffers a loss and reports this loss to the insurance company.

_____ 8. At a claim center, an insurance adjuster _____.
　A. examines a damaged vehicle and writes an estimate
　B. repairs the vehicle damage
　C. customizes vehicle features
　D. applies a new topcoat of paint to a vehicle

Copyright by Goodheart-Willcox Co., Inc. May not be reproduced or posted to a publicly accessible website.

_____ 9. *True or False?* If an insurance company has a direct repair agreement with a body shop, the person with a claim can take the damaged vehicle directly to that shop.

_____ 10. *True or False?* Replacement vehicle parts should be ordered three to five days after the vehicle is brought into the shop for repair.

_____ 11. To avoid delays, many shops _____ a damaged vehicle so all damage is discovered and required parts are available before the repairs begin.

_____ 12. Which type of part is made by the vehicle's manufacturer?
  A. Unlicensed.
  B. Aftermarket.
  C. OEM.
  D. All of the above.

_____ 13. OEM stands for _____.

_____ 14. _____ are made by a company other than the one that supplied the parts when the vehicle was manufactured.

_____ 15. *True or False?* Aftermarket parts usually cost more than OEM parts.

_____ 16. _____ are taken off total loss vehicles in a salvage yard.

_____ 17. *True or False?* Used parts are normally sold as individual parts with related components sold separately.

_____ 18. *True or False?* Used parts may have damage or they may have been previously repaired.

_____ 19. Which of the following can be obtained as a remanufactured part?
  A. Alloy wheel.
  B. Chrome bumper.
  C. Bumper cover.
  D. All of the above.

For questions 20–23, match the following part types with their descriptions.

_____ 20. Made by the vehicle manufacturer or by the same company that supplied the part to the factory when the vehicle was built.

_____ 21. Made by a company other than the one that supplied the parts when the vehicle was built.

_____ 22. Taken off total loss vehicles in a salvage yard.

_____ 23. Has been restored to like-new condition.

A. Remanufactured part
B. OEM/Original equipment manufactured part
C. Used part
D. Aftermarket part

Chapter 1 Introduction to Collision Repair

Name _____

_____ 24. The process of dismantling a vehicle to replace parts and access hidden damage is known as _____.

_____ 25. Structural repair on a vehicle demands a technician take proper _____ to pinpoint the damage.

_____ 26. With a vehicle immobilized, hookups are attached to a damaged area and structural components are _____ into proper alignment.

_____ 27. When only a portion of a damaged vehicle panel is replaced, the procedure is known as _____.

_____ 28. Lightly damaged vehicle panels can often be repaired, or _____.

_____ 29. *True or False?* A technician must find all of the damage if he/she intends to repair it properly.

_____ 30. When the metal of a vehicle body part has been moved to within _____" of its proper location, the remaining damage can be filled with plastic body filler.

_____ 31. Which of the following is a mechanical repair?
A. Replace an air conditioning system condenser.
B. Fix a fender.
C. Repair a quarter panel.
D. Straighten a frame rail.

_____ 32. *True or False?* After a collision, vehicle seat belts must also be checked for damage as part of the repair process.

_____ 33. *True or False?* Brake pads are the most vulnerable brake system parts in a collision.

_____ 34. Layers of paint designed to bind with bare metal, prevent rust, and fill scratches are known as _____.

_____ 35. Block sanding _____ the repair area, making it the same height as the surrounding paint.

_____ 36. *True or False?* The topcoat will hide any imperfections in the undercoat or body filler.

_____ 37. The _____ is sprayed over the undercoat and blended into the paint on adjacent vehicle panels.

_____ 38. *True or False?* Many late-model vehicles use a single-stage topcoat.

_____ 39. *True or False?* Some parts are painted off the vehicle and installed after the paint has cured.

40. Which is the last step in the collision repair sequence?
    A. Paint.
    B. Mechanical.
    C. Structural.
    D. Detail.

41. *True or False?* Stripes or decals are applied during the painting phase.

42. *True or False?* An independent body shop is connected to a nationwide chain of locations owned by a large corporation or organization.

43. Due to its size and ownership, a franchise body shop may be able to offer _____ for employees than an independent body shop.
    A. friendlier coworkers
    B. more benefits
    C. nicer customers
    D. less work

44. *True or False?* Dealership body shops often perform warranty work.

45. Which type of body shop focuses on restoration and custom work?
    A. Independent body shop.
    B. Franchise body shop.
    C. Dealership body shop.
    D. Specialty body shop.

46. *True or False?* In a production job, the employee performs the actual repairs.

47. _____ involve tasks that support the production jobs, such as writing estimates and obtaining parts.

48. A body technician is responsible for all of the following tasks *except*:
    A. decal application.
    B. panel alignment.
    C. plastic repair.
    D. dent repair.

49. A(n) _____ must be proficient in spray gun usage, block sanding, color matching, and buffing.

50. A(n) _____ technician diagnoses and repairs structural damage.

51. A structural technician must be able to perform all of the following tasks *except*:
    A. welding.
    B. color matching paint.
    C. hooking up pulling equipment to a vehicle.
    D. removing parts.

Chapter 1 Introduction to Collision Repair

Name _____

_____ 52. Which body shop employee repairs the air conditioning system of a vehicle?
    A. Refinish technician.
    B. Structural technician.
    C. General automotive repair technician.
    D. Detailer.

_____ 53. A(n) _____ cleans repaired vehicles before delivery back to the customer.

_____ 54. *True or False?* Combination technicians perform all types of repairs on damaged vehicles.

_____ 55. All body shop production jobs require _____.
    A. eye-hand coordination
    B. mechanical aptitude
    C. attention to detail
    D. All of the above.

_____ 56. The _____ is responsible for ordering parts, scheduling work, talking to customers, and evaluating their technicians' performance.

_____ 57. The customer usually encounters which body shop employee first?
    A. Estimator.
    B. Office manager.
    C. Shop foreman.
    D. Parts person.

_____ 58. *True or False?* An estimator must have extensive knowledge of vehicle repair in order to accurately compile a repair cost quote.

_____ 59. The _____ supervises the estimator and parts personnel, and is responsible for accepting payment from the customer.
    A. wrecker operator
    B. office manager
    C. detailer
    D. refinish technician

_____ 60. *True or False?* Part of the insurance adjuster's job is to maximize the cost of claims.

_____ 61. An automotive paint store employee, also called a(n) _____, sells paint and supplies to body shops.

_____ 62. Which body shop support position would remove hail dents?
    A. Paint jobber.
    B. Paintless dent repair technician.
    C. Auto recycler.
    D. Insurance adjuster.

Chapter 2

Safety

Learning Objectives

After studying this chapter, you will be able to:
- Recognize the hazards encountered in the collision repair shop.
- Demonstrate the proper use of the protective gear worn in the collision repair shop.
- Describe the safe working practices associated with the various tasks performed in the collision repair shop.
- Identify the hazardous wastes generated in the collision repair shop.
- Explain how the amount of waste generated in the shop can be minimized.
- Describe the role of government agencies in regulating collision repair shop wastes.

Carefully read Chapter 2 of the textbook and then answer the following questions in the space provided.

_____ 1. A hazard that can cause long-term health damage, usually from repeated or prolonged exposure to the hazard is known as a(n) _____.

_____ 2. A _____ is a chemical, commonly used in an auto body shop, which can cause chronic damage to the central nervous system.
   A. blood toxin
   B. neurotoxin
   C. liver toxin
   D. reproductive toxin

_____ 3. A(n) _____ hazard can cause an immediate (or within a few hours) reaction, even from a single exposure.

_____ 4. *True or False?* The acute reaction to isocyanates causes asthma-like symptoms.

5. Why should a technician make sure the back side of a vehicle panel is accessible before welding?

_____

_____

_____ 6. *True or False?* The heat of a fire is a more frequent cause of death than the smoke from a fire.

_____ 7. *True or False?* Never enter a smoke-filled room to put out a fire.

8. Which of the following elements does a fire require in order to burn?
   A. Combustible material.
   B. Heat.
   C. Oxygen.
   D. All of the above.

9. Wood, paper, plastic, or cloth fires are classified as _____ fires.

10. *True or False?* Type B fires should be smothered to remove the oxygen.

11. *True or False?* Water can be used successfully when attempting to put out a liquid fire.

12. Type C fires can involve one or all of the following items *except*:
    A. electrical motors.
    B. wiring.
    C. wood.
    D. electrical switches.

13. What items are involved in a fire classified as a Type D fire?

14. *True or False?* The most common type of multipurpose extinguisher is an A, B, C dry-chemical fire extinguisher.

15. A(n) _____ liquid is a liquid that evaporates readily when exposed to the air, producing extremely flammable vapors.

16. What do the letters *VOC* stand for?

17. When a volatile liquid is _____ or broken up into small droplets, the mist created presents an even greater explosion hazard than the liquid itself.
    A. atomized
    B. heated
    C. cooled
    D. mixed

18. The replacement of all the contaminated air in a confined area, such as a paint mixing room or paint storage room, by a shop ventilation system is called a(n) _____.

19. Fresh air, pushed into a spray booth from outside the building via an intake fan, is called _____.

20. When sanding an aluminum panel, a(n) _____ system should be used to help prevent an explosion.

21. The process of adding solvents to paint to dilute the paint into a sprayable consistency is called _____.
    A. deduction
    B. reduction
    C. induction
    D. conduction

Chapter 2 Safety

Name _____

_____ 22. Hazardous air pollutants (HAP) can enter the body through _____
_____   or _____.

23. What are the short-term and long-term health risks of HAP exposure?

_____
_____
_____
_____
_____

_____ 24. Sunlight causes a reaction between VOCs and _____ to make
photochemical smog and ground level ozone.
   A. carbon monoxide
   B. hydrogen
   C. nitrogen oxide
   D. All of the above.

25. How are VOCs in paint measured?

_____

_____ 26. The measure of the amount of paint that coats a surface compared
to the amount of paint that is actually sprayed is called _____.
   A. transfer efficiency
   B. sensitization
   C. oxidization
   D. paint overlap

_____ 27. *True or False?* A high-volume low-pressure (HVLP) spray gun requires
significantly more paint to refinish a vehicle than a standard spray gun.

_____ 28. Paint hardeners, or catalysts, contain a family of chemicals called _____.

29. What is *sensitization*?

_____
_____
_____
_____
_____

_____ 30. *True or False?* Most sensitized individuals cannot work in a collision
repair shop.

_____ 31. *True or False?* Carbon monoxide (CO) can be smelled.

_____ 32. When dealing with electrical devices, never do which of the following?
   A. Use electric tools or handle live extension cords while standing on wet floors.
   B. Operate several electrical devices from a single extension cord.
   C. Operate high-current devices, such as air compressors or welders, through extension cords.
   D. All of the above.

_____ 33. While working on a hybrid vehicle, a technician must remove or switch off the _____ to prevent current through the vehicle's high voltage system.
   A. compressor
   B. fuse box
   C. high-voltage disconnect
   D. actuator

_____ 34. In an auto body shop, the pounding sound waves produced by loud noises can damage _____.

35. What can happen to the unprotected skin from the light of MIG welding?

_____

36. Describe how to lift a heavy object from the floor.

_____
_____
_____
_____

_____ 37. A _____ is the inflammation of muscles or tendons caused by performing the same action again and again.
   A. repetitive stress injury
   B. fracture
   C. concussion
   D. None of the above.

_____ 38. When carrying an undeployed (live) air bag module, the module should be pointed _____ from the technician's body.

_____ 39. True or False? Safety glasses should be worn so they fit loosely against the forehead.

_____ 40. True or False? Safety glasses should be worn with a face shield because contaminants may be able to get around the shield.

_____ 41. True or False? If a metal particle flies into a technician's eye, rubbing the eye will knock the particle safely out.

_____ 42. Earmuffs and earplugs will _____ the harmful sound waves produced in a collision repair shop.
   A. heighten
   B. dampen
   C. increase
   D. not affect

Chapter 2 Safety

Name _____

_____ 43. A(n) _____ must be worn when dry sanding to protect the respiratory system.

44. Explain how to clean a cartridge-type respirator.

_____
_____
_____
_____
_____
_____
_____
_____

_____ 45. *True or False?* A cartridge respirator cannot remove isocyanates from the air and should never be used when spraying isocyanate-containing paint.

_____ 46. A(n) _____ provides the paint technician with clean, safe air from outside the spray area.

_____ 47. Due to harmful chemicals, wear _____ gloves when mixing or spraying paint, cleaning a spray gun, and handling solvents.

_____ 48. What shade should a MIG welding helmet have?
 A. #4.
 B. #6.
 C. #8.
 D. #10.

_____ 49. *True or False?* Canvas shoes provide superior protection from welding sparks.

_____ 50. When using an electric welding or cutting process, _____ the vehicle's negative battery cable.

51. What could happen if a welding cylinder falls over and breaks its valve?

_____
_____

_____ 52. *True or False?* If a hookup fails while a pull is being made, the chain may fly back toward the source of the pull.

53. How do you prevent fly back while making a tension repair?

_____
_____

54. List the protective equipment that should be worn when grinding or cutting.
_____
_____

55. List the protective gear that should be worn when spraying isocyanate-containing paint.
_____
_____

56. List the safety equipment to wear when painting with non-isocyanate paint.
_____
_____

_____ 57. *True or False?* Without adequate power ventilation, the isocyanates in paint may be present in the air for hours or days after the paint is sprayed.

58. According to guidelines from the Environmental Protection Agency (EPA), a collision repair shop has a "cradle-to-grave" responsibility for hazardous wastes. Explain what the term *cradle-to-grave* means for the shop.
_____
_____
_____
_____

_____ 59. Each collision repair shop is required to have an EPA _____ disposal number which allows the agency to track proper disposal of these harmful substances.
   A. metal waste
   B. paper waste
   C. hazardous waste
   D. food waste

_____ 60. *True or False?* The EPA National Rule limits the amount of metallic flakes that can be in various types of paint.

_____ 61. *True or False?* Ethyl acetate is an example of an exempt solvent.

_____ 62. The mission of the _____ is to make sure employers provide a safe and healthy work environment.

_____ 63. *True or False?* OSHA also regulates shop practices and shop equipment.

_____ 64. _____ laws require employers to inform workers about the hazardous chemicals in the workplace.

_____ 65. More detailed than a product label, the _____ lists the individual chemical components of a product and the hazards of each component as well as firefighting and clean-up procedures.

Name _____ Date _____ Class _____

Chapter 3

Vehicle Construction

Learning Objectives

After studying this chapter, you will be able to:
- Explain the different vehicle classifications.
- List the materials used to construct the automotive body.
- Identify the major assemblies of an automotive body.
- Summarize the vehicle production process.
- Describe vehicle safety ratings.

Carefully read Chapter 3 of the textbook and then answer the following questions in the space provided.

_____ 1. Vehicles are commonly classified by construction, as either _____ vehicles or _____ vehicles.

_____ 2. *True or False?* A full-frame vehicle has a body and separate frame.

_____ 3. List the steel/aluminum thickness range of vehicle full frames.

_____ 4. *True or False?* Aluminum frames weigh more than steel frames.

_____ 5. A vehicle constructed of sheet metal panels that are spot welded together to form one main body structure is classified as a _____ vehicle.
   A. multiple piece
   B. full-frame
   C. unibody
   D. flexible

6. When examining two vehicles of comparable size, which vehicle weighs more, a unibody or a full-frame?

   _____

_____ 7. The vehicle's _____ produces power and conveys this power to the drive wheels.

Copyright by Goodheart-Willcox Co., Inc. May not be reproduced or posted to a publicly accessible website.

21

_____ 8. In a rear-wheel-drive vehicle with a front-mounted engine positioned longitudinally, a driveshaft connects the transmission to the _____.
   A. differential
   B. engine
   C. front axles
   D. None of the above.

_____ 9. *True or False?* The engine is mounted longitudinally in all four-wheel-drive vehicles.

_____ 10. The electric motor alone does *not* power the vehicle in a _____ hybrid.
   A. parallel
   B. full
   C. series
   D. main

11. Identify the components of the four-wheel-drive system shown in the following illustration.

   A. _____
   B. _____
   C. _____
   D. _____
   E. _____
   F. _____
   G. _____
   H. _____

_____ 12. *True or False?* A minivan has a fabric or hard top that can be folded down to allow open-air driving.

13. What are some of the advantages of a crossover vehicle compared to a standard SUV?

_____

_____

_____

_____ 14. A 3/4 ton pickup can safely carry up to _____ pounds.
   A. 500
   B. 1000
   C. 1500
   D. 2000

Chapter 3 Vehicle Construction

Name _____

_____ 15. *True or False?* An alloy is a metallic substance that consists of one metal and at least one other alloying element.

16. List the three basic types of steel used in automobile construction.
_____
_____

17. List four alloying elements in mild steel.
_____
_____

18. What is the yield strength of mild steel?
_____
_____

_____ 19. Crowns, body lines and convolutions give mild steel body panels greater _____.

_____ 20. *True or False?* Heating and welding mild steel will cause the steel to lose strength.

_____ 21. All the following elements are alloying elements in both high-strength and mild steel *except*:
A. carbon.
B. silicon.
C. copper.
D. molybdenum.

_____ 22. *True or False?* Conventional high-strength steel is harder and more brittle than mild steel.

_____ 23. Which of the following steels is the strongest?
A. Mild steel.
B. High-strength, low-alloy steel.
C. Advanced high-strength steel.
D. Conventional high-strength steel.

24. What are microstructures in steel?
_____
_____

25. List two types of advanced high-strength steel.
_____
_____

_____ 26. Made of two layers of steel bonded together with a solid layer of adhesive, _____ steel is used on cowl and floor areas of some vehicles.

_____ 27. Which steel rolling process is used to make full-frame rails?
   A. Hot.
   B. Back.
   C. Foil.
   D. Cold.

_____ 28. *True or False?* Hot-rolled steel achieves a better surface finish than cold-rolled steel.

29. What does a zinc coating on steel prevent?
_____

_____ 30. Aluminum is only about _____ the weight of steel.

_____ 31. Aluminum alloys are divided into which two types?
   A. Wrought and flexible.
   B. Wrought and cast.
   C. Cast and rigid.
   D. Cast and semirigid.

_____ 32. *True or False?* Heat-treatable aluminum alloy is heated after forming for increased strength.

33. How are extruded parts formed?
_____
_____
_____

34. List the melting point of steel and aluminum.
_____
_____

_____ 35. When comparing aluminum to steel, which of the following characteristics is *false* about aluminum?
   A. More difficult to recycle.
   B. Does not change color when heated.
   C. Melts at a lower temperature.
   D. Transfers heat more readily.

_____ 36. *True or False?* Magnesium parts can be heated and welded.

For questions 37–41, match the following material types with their descriptions.

_____ 37. Contains boron as an alloying element.
_____ 38. Types include wrought and cast.
_____ 39. Has a carbon content of approximately 0.25%.
_____ 40. Can catch fire if heated or welded.
_____ 41. Reduces noise, vibration, and harshness.

A. Advanced high-strength steel
B. Mild steel
C. Laminated steel
D. Aluminum
E. Magnesium

Chapter 3 Vehicle Construction

Name _____

_____ 42. _____ plastic can be easily distorted and returned to its normal shape after the pressure is released.

_____ 43. Radiator fan shrouds and fender liners are generally made of _____ plastic.

_____ 44. *True or False?* The glass used in a unibody vehicle is designed to strengthen the unibody.

_____ 45. _____ glass is made of two layers of glass with a layer of plastic in between them.

_____ 46. A single layer type of glass that shatters into small pieces, or shards, when broken is called _____ glass.

47. Identify the outer body assemblies shown in the following image.

A. _____
B. _____
C. _____
D. _____

_____ 48. The _____ form the sides of a vehicle's body behind the side doors and surround the rear wheel openings.
A. fenders
B. quarter panels
C. sail panels
D. rocker panels

_____ 49. On a pickup truck, a(n) _____, made out of steel, opens to allow cargo to be loaded into the bed.

_____ 50. On a unibody vehicle design, the _____ holds the condenser and radiator and may also provide a mounting for the headlights.

_____ 51. The _____ is a single panel that connects the upper and lower vehicle frame rails.

_____ 52. *True or False?* The location of the strut tower determines front end alignment.

_____ 53. The A pillar is also called the _____.
A. windshield pillar
B. center pillar
C. rocker panel
D. sail panel

_____ 54. The _____ is the base of the passenger compartment and has reinforced areas to which the seat belt anchors are mounted.
   A. wheel house
   B. cowl
   C. floor pan
   D. rear pillar

_____ 55. A(n) _____ surrounds the rear wheel and is formed from an inner and outer panel.

_____ 56. On replaceable full frame sections, cross members may be _____ or _____ to side rails.

_____ 57. Sound deadener limits the _____ of the body panels.

58. During vehicle production, what manufacturing process uses spot welds and adhesives?
_____

_____ 59. *True or False?* A one-piece door opening panel on an extended cab pickup truck is an example of a laser welded part.

_____ 60. In hydroforming, a rectangular steel tube is formed into a complex shape using _____ pressure.

_____ 61. Some portions of a full frame may be C-shaped and other portions are _____.

_____ 62. _____ foam, which is molded to the internal shape of some vehicle cavities, adds strength and stiffens a vehicle part.
   A. Structural
   B. Collapsible
   C. Loose
   D. None of the above.

_____ 63. After performing a full-width frontal crash test, if the dummies have less than a(n) _____ chance of sustaining a life-threatening injury, the vehicle receives a 5-star front impact safety rating

_____ 64. Which vehicle safety test uses a 3000 pound sled to hit a vehicle?
   A. Full-width frontal.
   B. Offset frontal.
   C. Side.
   D. Roof strength.

_____ 65. *True or False?* In a roof strength rating test, a force-to-weight ratio of 4:1 or more earns the vehicle a poor rating.

Name _____ Date _____ Class _____

Chapter 4

Fundamentals of Collision Damage

Learning Objectives

After studying this chapter, you will be able to:
- Describe the various factors that will influence the way a vehicle will react in a collision.
- Identify bends, body lines, and crowns.
- Differentiate between direct and indirect damage.
- Explain the bend-versus-kink rule.
- Describe the types of collision damage found on full-frame and unibody vehicles.

Carefully read Chapter 4 of the textbook and then answer the following questions in the space provided.

_____ 1. _____ are formed by the crystalline pattern of the metal and alloy molecules.

_____ 2. What type of steel has a small, tight grain pattern?
   A. Mild steel.
   B. Drawing steel.
   C. High-strength steel.
   D. Deep drawing steel.

_____ 3. *True or False?* Differences in steel color in a bend or stretch translate into differences in the metal's properties in these areas.

4. Pressure is applied to a panel and distorts the metal. Pressure is released, and the metal springs back. Why?
   _____
   _____
   _____
   _____

_____ 5. What term refers to the ability of metal to be shaped?
   A. Plasticity.
   B. Elastic deformation.
   C. Plastic deformation.
   D. Elasticity.

6. Explain the difference between a *bend* and a *buckle* in a metal part.

_____

_____

_____

_____ 7. The permanent change in metal grain arrangement is _____.
   A. elastic deformation
   B. plastic deformation
   C. internal deformation
   D. unworking deformation

_____ 8. Both bends and buckles are formed by the _____ of the outer grains and the _____ of the inner grains.

_____ 9. *True or False?* Work hardening in severe bends, such as in body lines, weakens the metal panel.

_____ 10. Unirails—the frame rails on unibody vehicles—have many bends, or _____, in them that provide structural strength and energy management during a collision.

_____ 11. Which type of crown is described as highly curved and rigid?
   A. Low crown.
   B. High crown.
   C. Double crown.
   D. Reverse crown.

12. What factors determine the strength of the crown?

_____

_____

_____

_____ 13. *True or False?* Since most panels have a combination crown, it is useful to think of crown in terms of panels rather than areas.

_____ 14. Different types of crown require different _____ methods.

_____ 15. In collision repair, _____ can be thought of as the energy exchanged when two or more objects collide.

_____ 16. When dropped from the same height, which of the following balls has more force?
   A. Bowling ball.
   B. Ping pong ball.
   C. Baseball.
   D. Basketball.

_____ 17. The _____ is the minimum amount of force required to cause a permanent change in grain arrangement.

Chapter 4 Fundamentals of Collision Damage

Name _____

18. What happens when force strikes an unattached object?

_____

_____ 19. *True or False?* Force applied from one end is called longitudinal force.

_____ 20. Due to the bend(s) used to make a vehicle body line, applying longitudinal force causes _____ buckling at the body line.

_____ 21. Force applied from the side is called _____ force.
   A. rear
   B. longitudinal
   C. front
   D. lateral

_____ 22. Depending on the plasticity of the sheet metal, once the limit of stretch has been exceeded, the metal will _____.

_____ 23. Lateral force causes stretching in low-crown panels and collapse and arrowheads in _____ panels.

24. What factors determine the inertia of an object?

_____

_____

_____

_____ 25. *True or False?* To move an object, a force less than the inertia of the object must be applied.

_____ 26. Inertia damage is only found when a vehicle is moving with at least a moderate speed and comes to a(n) _____.
   A. steep incline
   B. gradual stop
   C. abrupt stop
   D. gradual incline

_____ 27. Which of the following factors determine how much damage a vehicle sustains in a collision?
   A. Target.
   B. Impact angle.
   C. Surface area of impact.
   D. All of the above.

_____ 28. When a vehicle hits a(n) _____ target, some of the collision force is used to overcome the inertia of the target, setting the target in motion.
   A. secondary
   B. unmovable
   C. movable
   D. None of the above.

_____ 29. *True or False?* When impact takes place at the center of a vehicle, the damage will be greater than if the impact takes place away from the center.

_____ 30. *True or False?* If the surface area of impact is large, the collision force is concentrated in a small area.

_____ 31. *True or False?* The frame or unirails are the parts of the vehicle most resistant to impact.

_____ 32. Damage that occurs at the point of contact in a collision is called ____.

_____ 33. ____ damage occurs away from the point of impact and is a result of the direct damage.

34. Why does paint chip off in an upset area?

_____
_____

35. List the names of the two types of buckles in an arrowhead.

_____

36. What is a *kink*?

_____
_____

_____ 37. *True or False?* A kinked high-strength steel vehicle panel must be replaced.

_____ 38. *True or False?* By looking at steel, a technician can determine if it is high-strength steel or mild steel.

_____ 39. *True or False?* A unibody vehicle is designed to collapse so the vehicle absorbs the impact forces and protects the passengers.

_____ 40. The ____ or frame step-in may act as pivot points in a front impact collision on a full-frame vehicle.

_____ 41. When the frame moves up, the change in frame height is called ____.

_____ 42. A change in vehicle frame length is called ____.
   A. sag
   B. mash
   C. sidesway
   D. twist

43. What type of damage occurs when a vehicle frame moves down?

_____

Chapter 4 Fundamentals of Collision Damage 31

Name _____

_____ 44. Lateral movement of a vehicle frame rail from a collision is known as ____.
A. sag
B. mash
C. sidesway
D. twist

_____ 45. After an impact, if the crossmembers and vehicle frame rails no longer meet at right angles, which type of damage has occurred?
A. Sag.
B. Mash.
C. Kickup.
D. Diamond.

_____ 46. If the frame rails under a pickup truck cab are not level with each other, the damage is called ____.

47. If lateral force strikes a full-frame pickup truck at a crossmember, what happens if the crossmember is strong?

_____
_____
_____

48. A rear impact to a full-frame pickup truck causes mash damage to the arch area of the frame. What happens to the leaf spring?

_____
_____
_____
_____
_____

_____ 49. *True or False?* Diamond damage cannot occur in a unibody vehicle.

50. What does a convolution do to collision force?

_____
_____
_____

_____ 51. All of the following features is evidence of unibody collision damage *except*:
A. buckles.
B. intact spot welds.
C. split sealer.
D. misaligned panels.

_____ 52. *True or False?* A unibody vehicle is designed to deflect collision forces.

_____ 53. When examining a unibody vehicle for front impact damage, start at the _____ and look for panel misalignment.

54. A front door drops when opened while examining a unibody vehicle for front impact damage. What has been damaged or moved back?

_____

_____ 55. As in a full-frame rollover, the amount of damage sustained by a unibody depends on the _____ the vehicle lands on and the type of rollover.

Name _____ Date _____ Class _____

Chapter 5

General-Purpose Tools and Equipment, Service Information

Learning Objectives

After studying this chapter, you will be able to:
- Identify and describe the general-purpose hand tools used by the collision repair technician.
- Identify and demonstrate the use of the power tools used by the collision repair technician.
- Describe the various types of shop equipment encountered in the collision repair shop.
- Demonstrate the safe use of floor jacks, jack stands, and lifts.
- Identify and explain the use of the service information found in the collision repair shop.

Carefully read Chapter 5 of the textbook and then answer the following questions in the space provided.

_____ 1. For a collision repair technician, acquiring a collection of _____ is usually a continuous process begun from a basic collection.

_____ 2. Professional-grade hand tools have a(n) _____ warranty.

_____ 3. Which type of tool is often sold by company representatives who make sales calls on automotive repair shops?
A. Professional.
B. Hobbyist.
C. Homeowner.
D. Recreational.

_____ 4. *True or False?* The most expensive tool is the best choice.

_____ 5. Do not use a tool to do a job for which it was not _____.

6. List the two basic types of toolboxes.
_____

7. Why should a toolbox contain at least one large drawer?
_____
_____

_____ 8. In order to prevent a toolbox from tipping, never open more than _____ drawer at a time.

9. Identify the types of wrenches shown in the following illustrations.

   A. _____
   B. _____
   C. _____
   D. _____

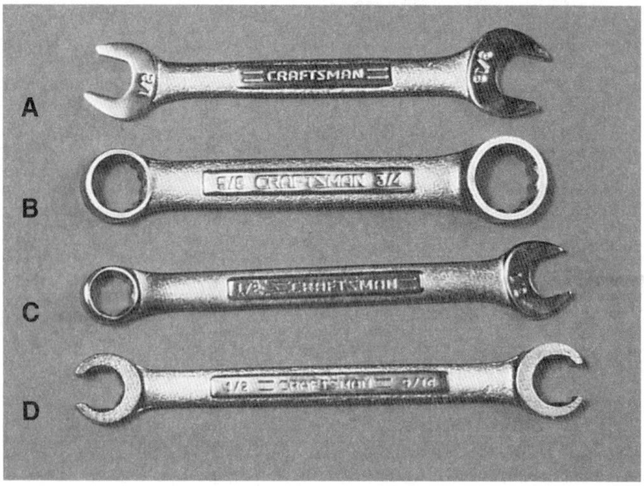

_____ 10. Which of the following wrenches is least likely to slip on fasteners?
   A. Tubing.
   B. Box-end 12-point.
   C. Open end.
   D. Box-end 6-point.

11. What is the advantage of an offset wrench?

   _____
   _____

_____ 12. _____ sockets are designed to reach a nut securing a protruding bolt.

_____ 13. *True or False?* Sockets available for socket wrenches include 1/4", 3/8", and 1/2" drive sizes.

_____ 14. Which of the following sizes is the largest Torx socket size?
   A. T15.
   B. T50.
   C. T30.
   D. T25.

_____ 15. *True or False?* When tightening a fastener, a ratchet lever is positioned so the ratchet locks when it is turned clockwise and releases when it is turned counterclockwise.

Chapter 5 General-Purpose Tools and Equipment, Service Information 35

Name _____

16. Identify the drive handles shown in the following illustrations.

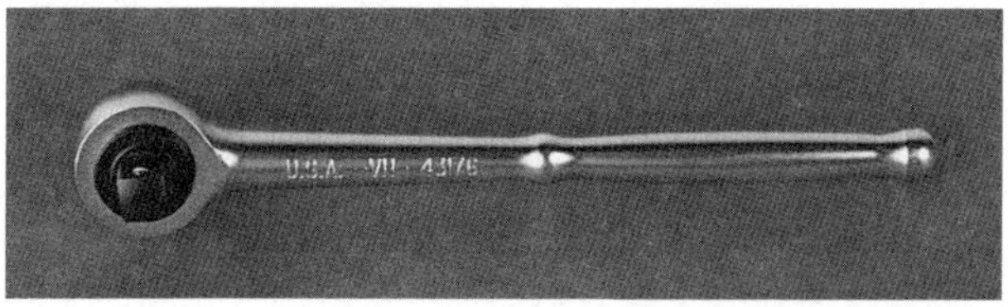

A

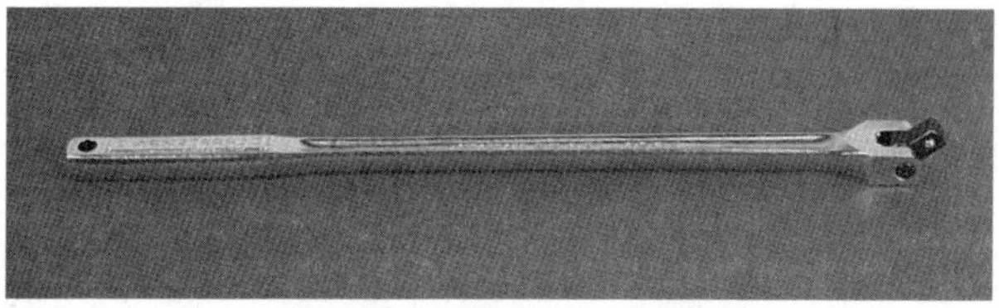

B

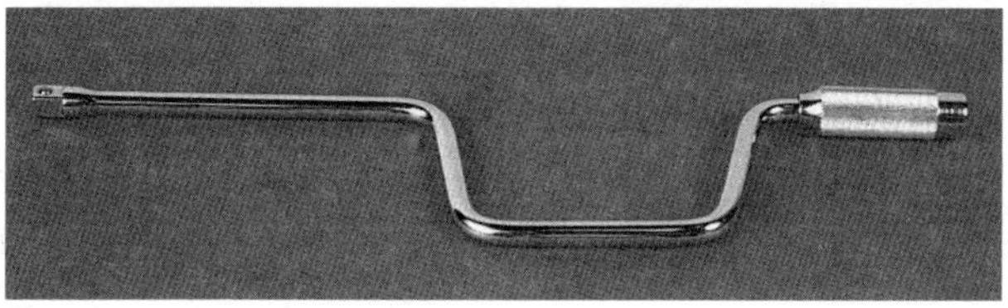

C

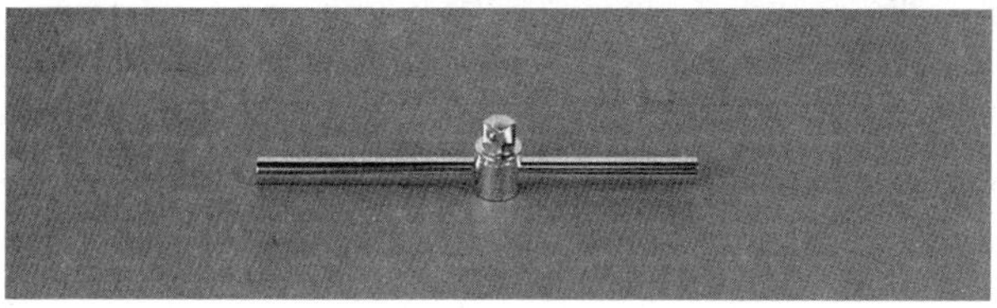

D

A. _____

B. _____

C. _____

D. _____

_____ 17. Torque wrenches are used to tighten bolts and nuts to exact _____.

18. A technician should use what tool to cut wire and remove cotter pins?

_____

_____ 19. Due to its firm grip, which type of plier would a technician use to remove rounded-off nuts and bolts?
   A. Diagonal cutting pliers.
   B. Slip joint pliers.
   C. Locking pliers.
   D. Snap ring pliers.

_____ 20. Locking pliers can be used as welding _____ when replacing panels.

_____ 21. _____ are used to secure a component in a hole or on a shaft.

22. Identify the pliers shown in the following illustrations.

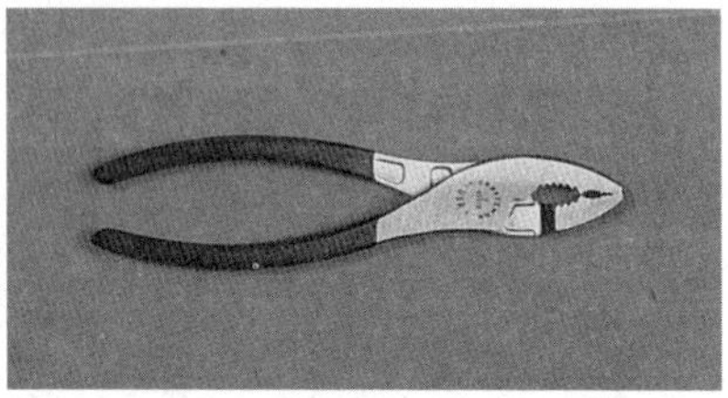

A.
B. _____
C. _____
D. _____
E. _____
F. _____

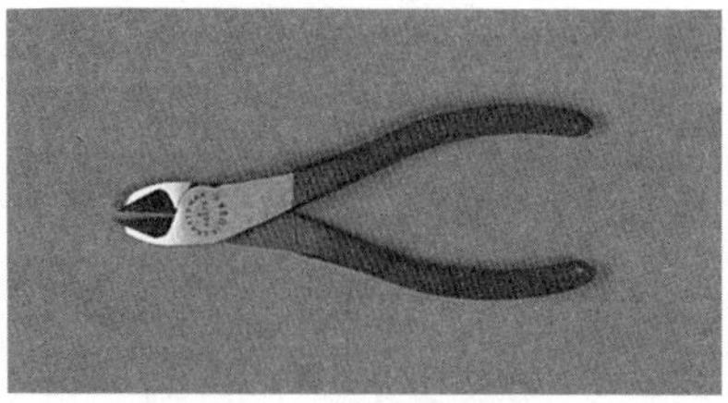

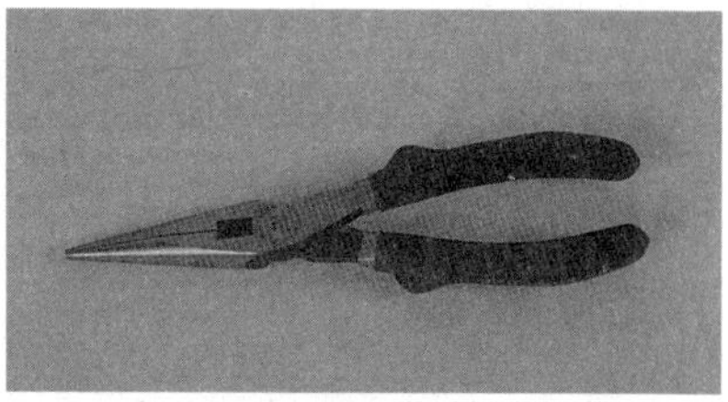

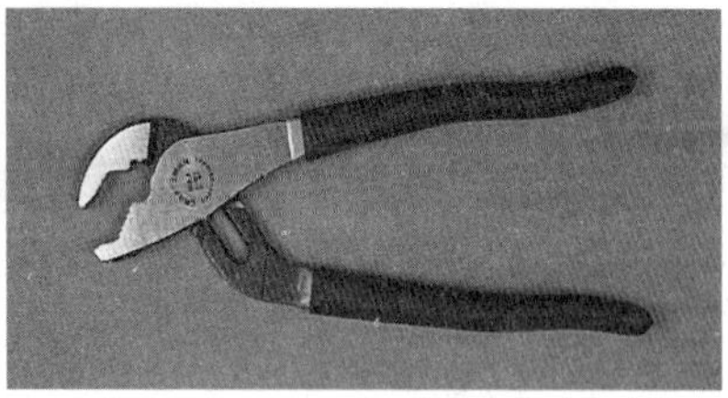

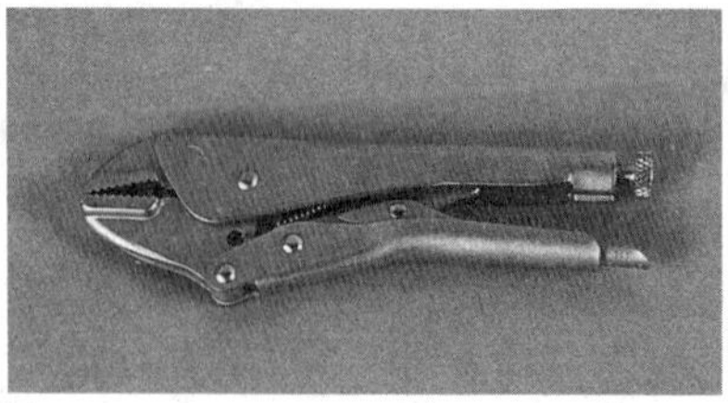

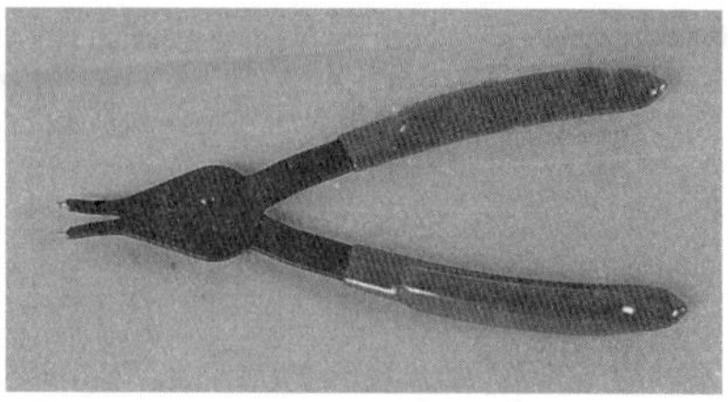

_____ 23. Make sure the screwdriver tip is properly _____ in the fastener before turning the screwdriver.

_____ 24. When prying, make sure the surface the bar rests against is _____ to prevent the bar from slipping.
   A. greasy
   B. slick
   C. breakable
   D. solid

25. What general-purpose tools are used for tapping and aligning?

_____

_____ 26. A(n) _____ hammer can be used to deliver a solid blow to the work surface without rebounding.

Chapter 5 General-Purpose Tools and Equipment, Service Information

Name _____

27. Identify the chisel types shown in the following illustrations.

   A. _____
   B. _____
   C. _____
   D. _____
   E. _____

   A

   B

   C

   D

   E

28. Explain how to align parts with an aligning punch tool.

   _____
   _____
   _____

_____ 29. Use a(n) _____ and _____ set to clean or repair the threads on nuts and bolts.

_____ 30. The hacksaw blade should be installed with the teeth pointing _____ from the handle.

_____ 31. _____ are useful for making quick, short—no longer than 6″—cuts in thin metal.

_____ 32. _____ are nonthreaded fasteners used to join metal together.

33. Why should a scraper have a blade that extends under the entire length of the handle?

   _____
   _____
   _____

34. When should a technician change the blade of a utility knife?

_____

_____ 35. Power tools either _____ operations or allow a technician to make repairs that would be impossible with hand tools.

36. How can a technician distinguish an impact socket from a non-impact socket?

_____

_____ 37. *True or False?* In general, a 1/4″-drive air ratchet is more powerful than a 3/8″-drive ratchet.

_____ 38. Variable speed means that the farther the trigger is pulled, the _____ the drill turns.

_____ 39. A small hole drilled in a panel to help in the alignment of a larger drill bit is called a(n) _____.

_____ 40. Which of the following items is considered shop equipment generally supplied by the collision repair shop?
    A. Floor jack.
    B. Compressed air system.
    C. Welding equipment.
    D. All of the above.

41. List the components of a compressed air system.

_____

_____

_____ 42. Which of the following is *not* a lift point for a vehicle?
    A. Crossmember.
    B. Unirail.
    C. Frame rail.
    D. Drive train.

_____ 43. *True or False?* A technician can safely work under a vehicle supported by only a floor jack.

_____ 44. The main drawback to a(n) _____ lift is that the cylinder obstructs the underside of the vehicle.

45. When raising a vehicle on a four-post lift, what should a technician do if the vehicle wobbles?

_____

_____

_____ 46. Which of the following lifts is the easiest to use, as a vehicle is simply driven onto the ramps?
    A. Four-post lift.
    B. Two-post lift.
    C. Below-ground lift.
    D. Ceiling lift.

Chapter 5 General-Purpose Tools and Equipment, Service Information

Name _____

_____ 47. A(n) _____ has wheels, allowing a technician to lie on it and roll under a vehicle to work.

_____ 48. *True or False?* The 110V MIG welder is best suited for joining structural and nonstructural sheet metal panels.

49. Which voltage of MIG welder should be used to weld full frames?

_____

_____ 50. How does a vehicle's computer warn the driver of a detected problem?
A. Malfunction indicator lamp.
B. Malfunction indicator alarm.
C. Both A and B.
D. Neither A nor B.

_____ 51. A(n) _____ must be used to read diagnostic trouble codes.

_____ 52. *True or False?* Vehicle computers do not store trouble codes in their memory.

_____ 53. _____ includes instructions for repairing or replacing damaged components, as well as information on refinishing.

_____ 54. Which of the following service information is *least likely* to be provided by the vehicle manufacturer?
A. Electrical repair.
B. Body repair.
C. Mechanical repair.
D. Troubleshooting procedures.

_____ 55. Which of the following is a source for collision repair information?
A. ICAR.
B. Tech Cor.
C. Paint manufacturers.
D. All of the above.

Name _____ Date _____ Class _____

Chapter 6

Fasteners

Learning Objectives

After studying this chapter, you will be able to:
- Explain galvanic corrosion.
- List the types of threaded fasteners.
- Understand torque-to-yield bolts.
- Recognize a welded nut or stud.
- Use liquid threadlock and anti-seize lubricant.
- List the types of screws.
- Understand the types of rivets.
- Use adhesives.
- List the types of plastic fasteners.

Carefully read Chapter 6 of the textbook and then answer the following questions in the space provided.

_____ 1. *True or False?* Galvanic corrosion occurs when two different types of metal come into direct contact with each other.

_____ 2. _____ coated fasteners are single-use and must be replaced if removed.

_____ 3. *True or False?* An isolation barrier is used to prevent direct contact between the aluminum and steel.

_____ 4. All of the following metals can be found on a vehicle *except*:
  A. aluminum.
  B. steel.
  C. magnesium.
  D. All of the above can be found on vehicles.

_____ 5. Threaded fasteners are being discussed. Technician A says bolts are threaded internally. Technician B says nuts are threaded internally. Who is correct?
  A. A only.
  B. B only.
  C. Both A and B.
  D. Neither A nor B.

_____ 6. _____ screws are used to temporarily join panels when bonding or welding.

Copyright by Goodheart-Willcox Co., Inc. May not be reproduced or posted to a publicly accessible website.

41

7. Which of the following factors must be considered when describing bolts?
   A. Diameter.
   B. Weight.
   C. Color.
   D. All of the above.

8. *True or False?* Bolt diameter is seldom measured in fractions of an inch.

9. Which of the following units of measurement is used to measure torque?
   A. Inch-pounds.
   B. Foot-pounds.
   C. Newton-meters.
   D. All of the above.

10. What are torque-to-yield bolts?

11. Identify the nuts shown in the following illustrations.
    A.
    B.
    C.
    D.
    E.
    F.

12. What is inserted through the castles of a castellated nut and a hole in the bolt or shaft to prevent loosening?
    A. Snap ring.
    B. Cotter pin.
    C. Rivet.
    D. Snap pin.

13. *True or False?* Self-locking nuts are designed to resist loosening caused by vibration.

14. *True or False?* J-clips may need to be held with a wrench as a bolt or screw is threaded into them.

15. What kind of nut is held in place by metal framework?

16. Washers increase the _____ of a bolt.

Chapter 6 Fasteners

Name _____

_____ 17. Washers are being discussed. Technician A says that washers can decrease the number of bolt threads that get threaded into a nut. Technician B says that washers have no effect on the number of bolt threads that get threaded into a nut. Who is correct?
A. A only.
B. B only.
C. Both A and B.
D. Neither A nor B.

_____ 18. A conventional flat washer can be used under a nut to prevent _____ the contact surface.

_____ 19. *True or False?* A stud is the non-threaded part of a bolt that is welded to a structure.

20. What is done if a damaged fastener cannot be turned?

_____

_____

_____

_____ 21. Trim screws often have a(n) _____ washer that fits into a beveled hole.

_____ 22. A sheet metal screw uses a _____ point to cut threads as it's screwed into sheet metal.
A. hexagonal
B. wide
C. fluted
D. All of the above.

_____ 23. *True or False?* Corrosion may prevent the removal of a screw.

_____ 24. When using an oxyacetylene torch to loosen a seized fastener, use a torch with a small tip and a(n) _____ flame to heat the nut evenly.

_____ 25. *True or False?* When using an air chisel on a seized bolt, the intent is to cut off the bolt.

_____ 26. Technician A says a 6-point socket provides better grip than a 12-point socket. Technician B says if a fastener is rounded, use locking pliers to grip it. Who is correct?
A. A only.
B. B only.
C. Both A and B.
D. Neither A nor B.

27. On a bolt, what is used to measure the number of threads per inch, or pitch if the bolt is metric?

_____

_____ 28. *True or False?* One method of thread repair uses a helical insert, a patented coil of wire that forms new threads as the bolt is inserted.

_____ 29. *True or False?* When replacing fasteners, it is important that the new fastener is equal in strength to the old fastener.

_____ 30. When a bolt is broken off flush with the surface during removal, what can be used to remove the broken portion remaining in the hole?
   A. Tap.
   B. Screw extractor.
   C. Bolt grabber.
   D. Bolt pry.

_____ 31. Some types of _____ can be applied to threaded fasteners to prevent loosening due to vibration but still allow for easy removal if disassembly is required.

_____ 32. *True or False?* Liquid threadlock is applied to the entire threaded portion of the fastener.

33. What does it mean when a bolt is said to be cold welded?

_____
_____
_____
_____

_____ 34. What can be applied to a bolt to prevent it from becoming cold welded to the connecting metal?

_____

_____ 35. What type of non-threaded fastener is used to prevent endwise movement of cylindrical parts and shafts?
   A. Cotter pin.
   B. Snap rings.
   C. Piston pin.
   D. None of the above.

36. What is a retaining ring and how does it work?

_____
_____
_____

_____ 37. *True or False?* Typically, when a rivet head is sheared off by a collision, the rivet hole in the frame is not damaged.

38. What type of rivet has a partially hollow nose?

_____

Chapter 6 Fasteners

Name _____

_____ 39. Rivets are being discussed. Technician A says that the use of blind rivets requires access to the inner panel only. Technician B says you only need access to the outer panel when using self-piercing rivets. Who is correct?
A. A only.
B. B only.
C. Both A and B.
D. Neither A nor B.

_____ 40. Which of the following can be used on aluminum panels?
A. Blind rivets.
B. Solid rivets.
C. Self-piercing rivets.
D. All of the above.

_____ 41. Technician A says a higher temperature shortens an adhesive's cure time. Technician B says cure time is the number of hours required for the applied adhesive to gain full strength. Who is correct?
A. A only.
B. B only.
C. Both A and B.
D. Neither A nor B.

_____ 42. *True or False?* Glass cannot be installed on a vehicle using adhesive.

_____ 43. The _____ is the number of minutes that a panel can be adjusted slightly for alignment before the adhesive has cured too much.

44. In weld bonding, what type of welder is typically used?
_____

45. What type of panel is commonly weld bonded?
_____

_____ 46. *True or False?* Hybrid bonding has the advantage of the full length bond of adhesive reducing noise and vibration coupled with the added strength of welds or rivets.

_____ 47. Which of the following is *least likely* to be held together with plastic fasteners?
A. Hood hinges.
B. Interior trim.
C. Door panel.
D. Bumper cover.

48. How is a plastic rivet removed?

_____

_____

_____

_____ 49. Snap brackets are _____ to the body of the vehicle.

_____ 50. *True or False?* Plastic clips with damaged mounting tabs must be replaced.

Name _____ Date _____ Class _____

Chapter 7

Welding and Cutting

Learning Objectives

After studying this chapter, you will be able to:
- List the safety precautions required for welding.
- Describe the types of welds used in the collision repair shop.
- Explain the types of joints encountered during panel repair or replacement.
- Explain how to set up a MIG welding machine.
- Recognize MIG welding variables and explain how to control them.
- Explain how to make various types of welds using a MIG welding machine.
- Understand the differences between aluminum and steel welding.
- Recognize two types of resistance spot welding and know when each is used.
- Explain how to use a plasma cutter and cutting torch.

Carefully read Chapter 7 of the textbook and then answer the following questions in the space provided.

_____ 1. _____ produces sparks, heat, and bright light.

2. When welding galvanized metal, what happens to the galvanized coating?
_____

_____ 3. When MIG welding, a technician should wear a helmet with a #_____ shade.

_____ 4. *True or False?* Never look at the MIG welding arc without the proper shaded vision protection.

5. Before welding on a vehicle, what should a technician do to the negative battery cable?
_____
_____

_____ 6. A spark that does not bounce off of a plastic welding blanket may _____ through the blanket, damaging the vehicle.

7. Why is it a good practice to stop welding one hour before the end of work time?
_____
_____
_____

_____ 8. Oxygen, acetylene, and other welding-gas cylinders should be chained to a cart or wall to prevent _____.

_____ 9. All of the following are basic types of welds used in a collision repair shop *except*:
   A. Stitch.
   B. Needle.
   C. Plug.
   D. Spot.

_____ 10. Which type of weld is made by welding around the perimeter of a hole?
   A. Stitch.
   B. Spot.
   C. Plug.
   D. Continuous.

_____ 11. Which of the following welding positions is the most difficult?
   A. Flat.
   B. Horizontal.
   C. Vertical.
   D. Overhead.

12. Identify the joint types shown in the following illustration.
   A. _____
   B. _____
   C. _____
   D. _____

Chapter 7 Welding and Cutting

Name _____

_____ 13. _____ occurs when the heat from the weld melts away the base metal.

_____ 14. Remove the paint from panels to be welded by sanding with _____.
   A. a grinder
   B. 80-grit sandpaper
   C. 24-grit grinding disk
   D. 36-grit sandpaper

_____ 15. The area of molten wire and metal created by the heat of the electric arc is called the _____.

16. What are the two methods of melting electrode wire in a MIG welder?

_____

_____ 17. MIG welding is being discussed. Technician A says that in short circuit MIG welding, the arc switches on and off 100 times per second. Technician B says that in pulsed MIG welding, there are anywhere between 30 and 400 pulses per second. Who is correct?
   A. A only.
   B. B only.
   C. Both A and B.
   D. Neither A nor B.

_____ 18. The inert gas in metal inert gas (MIG) welding is _____.
   A. CO_2
   B. O_2
   C. argon
   D. neon

19. What does the work clamp on a MIG welder do?

_____

_____

20. Identify the parts of the MIG welder shown in the following illustration.

A. _____
B. _____
C. _____
D. _____
E. _____
F. _____
G. _____
H. _____
I. _____

_____ 21. The _____ in the drive roller must be the proper size for the diameter of the welding wire being used.

22. What is the function of deoxidizer in MIG welding wire?

_____
_____
_____

23. The label on a MIG welding wire says AWS ER70S-6. What does the 70 mean?

_____

24. The label on a MIG welding wire says AWS ER70S-6. What do the letters AWS mean?

_____

_____ 25. *True or False?* MIG welding machines used for collision repair of steel use a mixture of 13% argon and 87% carbon dioxide.

_____ 26. When setting up the MIG welder shielding gas, the regulator gauge should be set at _____ cfm or _____ psi.

_____ 27. The size of a MIG welding arc is determined by which of the following factors?
   A. Voltage.
   B. Pressure.
   C. Wire speed.
   D. Shielding gas.

Chapter 7 Welding and Cutting

Name _____

_____ 28. The wire speed setting is determined by the thickness of the metal to be welded and the _____ of the welding electrode.

_____ 29. A _____ screws into the end of a welding gun and establishes a good electrical connection between the gun and the welding wire.
   A. motor
   B. nozzle
   C. contact tube
   D. fan

_____ 30. *True or False?* Flux-cored wire requires shielding gas.

31. A MIG welder has a 40% duty cycle rating. In a period of 10 minutes, how many minutes can be spent welding?

_____

_____ 32. _____ is the hardened remains of molten metal that is thrown about during the welding process.

_____ 33. _____ is the depth of melting during a weld.

_____ 34. As a general rule, 1 amp is required for each _____" of metal thickness.

35. List the MIG welding variables that the technician can influence.

_____

_____

_____ 36. The gun orientation should be at a 90° angle to the workpiece when making what type of weld?
   A. Plug.
   B. Lap.
   C. Butt.
   D. Flange.

_____ 37. _____ is often referred to as the backhand method of welding.

_____ 38. Which travel direction is best suited for welding sheet metal?

_____ 39. The length of wire protruding from the welding gun's contact tube to the panel is called the _____.

_____ 40. Vertical up welds are made at _____ wire speed/amps and volts/heat settings than vertical down welds.

_____ 41. Prior to welding, which of the following steps should a technician do first?
   A. Check the thickness of the metal.
   B. Check for set up values inside the welder.
   C. Dial in set up values.
   D. Attach the work clamp.

_____ 42. Which of the following problems causes spatter?
   A. Amps too low.
   B. Shielding gas flow too high.
   C. Travel speed too slow.
   D. Heat too high.

_____ 43. The most common plug weld hole size in collision repair is _____.

44. Explain how to plug weld three pieces of metal together.

_____
_____
_____
_____
_____
_____

_____ 45. When using a stitch weld on a butt joint with an insert, the gap between the panels should be _____ times the thickness of the panel for best results.

46. Explain how to make a butt joint spot weld.

_____
_____
_____
_____

_____ 47. *True or False?* Aluminum changes color when heated.

_____ 48. *True or False?* Welding gun travel speed for aluminum should be slower than that for welding steel.

_____ 49. *True or False?* The larger heat affected zone (HAZ) that is created when MIG brazing protects the advanced high-strength steel from damage.

50. What is the tensile strength of MIG brazing wire?

_____

Chapter 7 Welding and Cutting

Name _____

_____ 51. What is the usual wire diameter selected for MIG brazing when using 100% argon gas?
   A. 0.010".
   B. 0.020".
   C. 0.030".
   D. 0.040".

_____ 52. _____ is contamination on a weld.
   A. Sag
   B. Slag
   C. Melt
   D. Bog

_____ 53. *True or False?* Compared to MIG welding, flux-colored arc welding (FCAW) produces more spatter and the welds have more slag.

_____ 54. In a tungsten inert gas (TIG) welder, the foot pedal or switch also controls the amount of _____ flowing through the tungsten electrode in the torch.

_____ 55. *True or False?* TIG welds should only be made on parts that have been removed from the vehicle.

56. List the four factors that determine the strength of a resistance spot weld.

_____

_____

57. What happens to the diameter of the electrode ends of a squeeze-type resistance spot welding (STRSW) machine after many welds have been made?

_____

_____

58. How many replacement welds should be made in a panel if the manufacturer recommends 1.3 replacement welds per factory spot weld and the panel has 30 factory spot welds?

_____

For questions 59–62, match the following welding processes with their descriptions.

_____ 59. Consumable filler material is added to weld by hand.

_____ 60. Process used to produce factory welds.

_____ 61. Most common type of welding performed in a collision repair facility.

_____ 62. Uses flux-filled welding wire.

A. MIG welding
B. Flux-cored arc welding
C. Tungsten inert gas welding
D. Resistance welding

_____ 63. Which of the following determines the amount of heat produced by an oxyacetylene torch?
A. Torch tip opening size.
B. Oxygen regulator setting.
C. Acetylene regulator setting.
D. Amount of welding gas in the cylinders.

_____ 64. When adjusting the gas regulators on an oxyacetylene torch, open the _____ tank valve approximately one-quarter to one-half turn.

_____ 65. *True or False?* An oxyacetylene torch can be lit with a cigarette lighter or another open flame.

_____ 66. A(n) _____ tip on an oxyacetylene torch is often used to heat steel during frame repair.

_____ 67. In order to use a 1000°F heat crayon, the metal must be heated to at least _____ first.

_____ 68. *True or False?* Since brazed joints are relatively strong, brazing can be used for any type of vehicle structural weld.

_____ 69. When cutting with a plasma cutter, travel speed is determined by _____.

_____ 70. *True or False?* An oxyacetylene torch can be used to rough cut high-strength steel and cut off rusted bolts from a vehicle.

Name _____ Date _____ Class _____

Chapter 8

Nonstructural Repair Tools, Equipment, and Materials

Learning Objectives

After studying this chapter, you will be able to:
- Describe the hand tools used in nonstructural panel repair.
- Identify the various power tools used when repairing or replacing nonstructural panels.
- Explain why filler is used and describe the different types of filler available.
- Describe the characteristics of various types of sandpaper.
- Explain the potential problems of cross contamination when repairing aluminum.

Carefully read Chapter 8 of the textbook and then answer the following questions in the space provided.

_____ 1. Which of the following procedures is *not* involved in repairing a damaged nonstructural vehicle panel to the proper contour?
   A. Pulling or pushing.
   B. Melting the whole panel down.
   C. Filling the low areas.
   D. Sanding.

_____ 2. On a body hammer with two working faces, when the pointed end is used, the hammer is called a(n) _____ hammer.

_____ 3. *True or False?* A large hammer face is useful on a low-crown surface.

_____ 4. *True or False?* A technician should use a circular face body hammer when attempting to straighten a body line.

_____ 5. The _____ of a hammer head determines the amount of force it will produce.
   A. texture
   B. color
   C. shape
   D. weight

_____ 6. A sharp pick body hammer can _____ a vehicle panel if used with a careless amount of force.
   A. dismantle
   B. crush
   C. pierce
   D. melt

_____ 7. Increasing the surface area of metal during a repair is called _____.

_____ 8. Decreasing the surface area of metal during a repair is referred to as _____.

_____ 9. What type of chisel hammer is used to tap down high spots in a large diameter body line?
   A. Sharp.
   B. Blunt.
   C. Serrated.
   D. None of the above.

10. Describe the face of a crown hammer.

_____

_____

_____

_____ 11. The _____ is the outer panel of a vehicle door.

12. How does tapping with a body hammer differ from driving a nail with a hammer?

_____

_____

_____

_____ 13. *True or False?* A technician placing his/her hand at the end of a hammer handle will deliver the maximum force per blow.

_____ 14. The hammer face should match the _____ contour of the panel as closely as possible.

15. Explain how to remove rust from the face of a body hammer.

_____

_____

_____

16. What happens to a hammer face if the hammer is used to hit something harder than the hammer face?

_____

_____ 17. _____ can be used to hit vehicle panel damage either directly or indirectly.
   A. Dollies
   B. Sandpaper
   C. Air saws
   D. Friction jacks

Chapter 8 Nonstructural Repair Tools, Equipment, and Materials

Name _____

18. How is a dolly used indirectly when repairing a vehicle panel?

_____

_____

_____

_____ 19. Which of the following items is *not* a type of dolly used in auto collision repair?
   A. Heel dolly.
   B. Comma dolly.
   C. Bear dolly.
   D. Spoon dolly.

20. Identify the dolly types shown below.

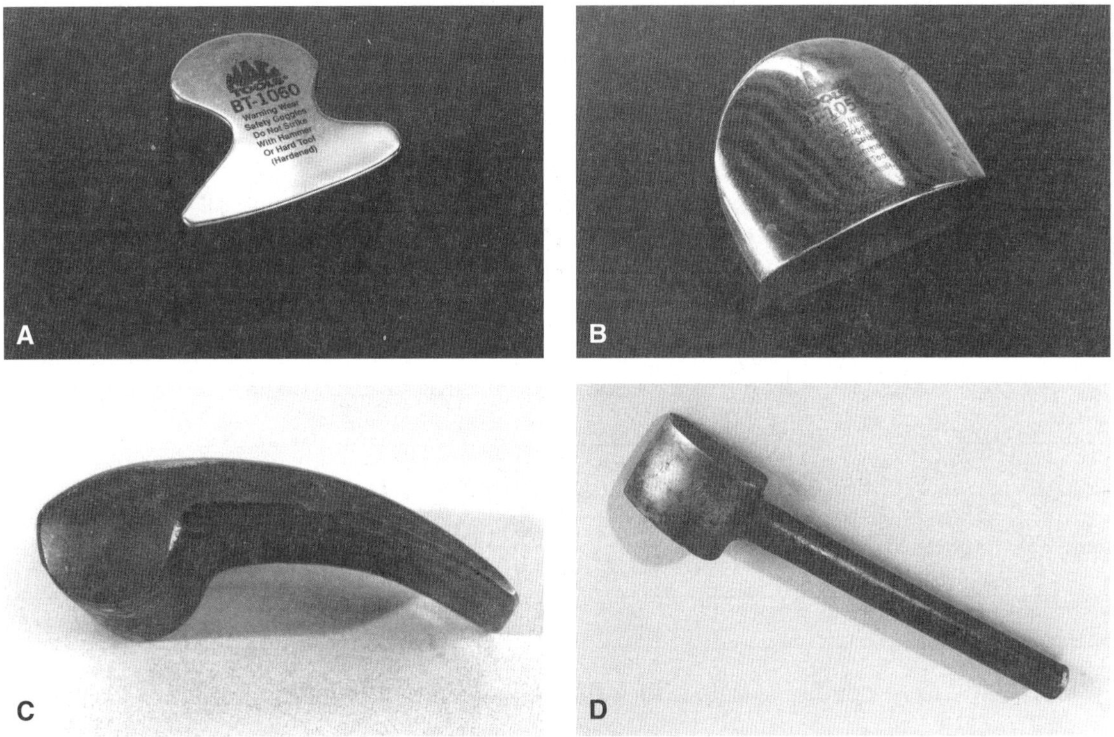

A. _____
B. _____
C. _____
D. _____

_____ 21. A(n) _____ is used to spread out the force of a hammer blow.

_____ 22. When in use, pry picks must have a surface to rest against for _____.

23. When removing damage with a pick that is inserted into a vehicle panel, how can a technician determine the pick location on the damage?

_____
_____
_____
_____

_____ 24. A deep curved hook attachment on a(n) _____ is designed to reach around obstructions and adjust vehicle panels or raise damage.

For questions 25–28, match the following body hammer types with their descriptions.

_____ 25. Features a smooth, round hammer face.

_____ 26. Features various attachments to align or raise vehicle panel damage.

_____ 27. Features a serrated hammer face.

_____ 28. Features two crowned surfaces, one crowned in each direction.

A. Shrinking hammer
B. Slide hammer
C. Stretching hammer
D. Crown hammer

_____ 29. *True or False?* When using a friction jack, installing the friction jack head on the rod so it moves away from the short leg of the rod will pull two panels together.

30. How does a technician set up an alignment bar?

_____
_____
_____
_____
_____
_____

31. List the tools used to remove a vehicle door with welded on hinges and a spring.

_____

32. Which direction should a technician push on a door adjust bar to lower the rear of the door?

_____

33. Why place masking tape over the working edge of a body chisel?

_____
_____
_____

_____ 34. A metal file should always be moved _____ to the crown of the vehicle panel when used to identify high and low areas.

_____ 35. *True or False?* Use long strokes when using a nonadjustable body file on a double crown vehicle panel.

Chapter 8 Nonstructural Repair Tools, Equipment, and Materials   59

Name _____

_____ 36. A damaged vehicle panel with good access to the _____ of the panel is an ideal place to use a bull's-eye pick.
A. inside
B. outside
C. edges
D. top

37. What is a *flange* on a vehicle panel?
_____
_____

38. Why can a hand punch tool only be used to make holes near the edge of a panel?
_____
_____

39. What happens if a technician mixes body filler on a dirty board?
_____
_____
_____

_____ 40. A filler _____ with a ragged or torn edge should be replaced to prevent any irregularities transferring to the filler.

_____ 41. Rough sanding removes excess filler with _____-grit sandpaper.

42. Rough sanding is complete when what appears all around the filler?
_____

_____ 43. It is important that the filler be sanded to exactly the same _____ as the surrounding metal.

44. What type of surface is required on a sanding tool used for final sanding?
_____

_____ 45. *True or False?* The sandpaper on a sanding tool should be installed tightly.

_____ 46. *True or False?* A long board should be held perpendicular to the crown of a panel and moved so it sands against the crown.

_____ 47. If a technician wanted to sand a reverse crown surface on a vehicle panel, sandpaper should be installed on the curved side of a(n) _____.

48. What are the two main problems that a technician faces when repairing a damaged vehicle body line with filler?
_____
_____
_____

_____ 49. Anchors embedded in the floor of a collision repair shop to which a power post is chained are known as _____.

50. How much of a grinder disk should contact a vehicle panel during use?

_____

_____

_____ 51. Two passes with a grinder over the same area of metal—one pass from right to left and the other from left to right—is called _____.
   A. looping
   B. buffing
   C. cross cutting
   D. dragging

_____ 52. *True or False?* An air file is used in the same way as a long board sander.

_____ 53. *True or False?* When using a spot weld cutter, always keep the cutter bit at a right angle to the panel.

_____ 54. Which of the following tools can cut holes along or crimp a step at the edge of a vehicle panel?
   A. Air punch/flanger.
   B. Air chisel.
   C. Cut-off tool.
   D. Electric shears.

_____ 55. A(n) _____ can be described as simply a power hacksaw.
   A. sanding block
   B. grinder
   C. impact wrench
   D. air saw

56. The working area of a door skin replacement tool consists of what two parts?

_____

_____

_____ 57. *True or False?* A draw pin welder can be used to shrink stretched areas of a vehicle panel.

_____ 58. A weld-on plate should *not* be welded across a(n) _____, as this must be allowed to lengthen and move during the pull.

59. When using a hydraulic power set, what happens if the base is weaker than the damage?

_____

_____

60. What product is used to fill shallow dents in vehicle panels that cannot be removed by straightening techniques?

_____

_____ 61. *True or False?* Lightweight body filler is waterproof.

Chapter 8 Nonstructural Repair Tools, Equipment, and Materials

Name _____

_____ 62. *True or False?* Heavyweight body fillers cure hard on the surface to produce a smooth, tack-free surface.

_____ 63. _____ is available in disk and sheet form and is rated by abrasive grit size.
A. Wire
B. Body filler
C. Sandpaper
D. Adhesive

64. What is the difference between *open-coat* and *closed-coat* sandpaper?

_____
_____
_____
_____
_____

_____ 65. *True or False?* A steel dolly used on an aluminum panel can also be used on a steel panel.

66. What is the major advantage of using adhesives over welding when replacing a vehicle panel?

_____
_____
_____

Name _____ Date _____ Class _____

Chapter 9

Nonstructural Panel Repair

Learning Objectives

After studying this chapter, you will be able to:
- Describe the individual steps in the nonstructural panel repair process.
- Explain how to rough out specific types of nonstructural damage.
- Demonstrate the metal finishing process.
- Use body filler to restore panel contour.
- Describe the considerations that must be taken into account when repairing aluminum panels.

Carefully read Chapter 9 of the textbook and then answer the following questions in the space provided.

_____ 1. After squeezing an empty soda can from the sides, the undamaged, flattened area held out of position between the arrowhead buckles is referred to as _____.

_____ 2. *True or False?* If buckles in a panel are released, the displaced metal will return to its proper position.

_____ 3. A(n) _____, or an increase in surface area, means that the grains in the metal have been thinned, flattened, and elongated.

_____ 4. A(n) _____, or a reduction in surface area, means that the grains in the metal have been shortened, thickened, and bunched together.

5. List three ways to locate a high and/or low spot of damage on a vehicle panel.
   _____
   _____

_____ 6. In general, for a nonstructural vehicle panel repair, if the cost of repair is more than _____% of the replacement cost, the damaged part is replaced.

7. What is a *double fold*?
   _____
   _____
   _____

_____ 8. *True or False?* A technician who can analyze and plan well will prevent wasteful backtracking when making repairs.

_____ 9. Which of the following steps is *not* important to consider when planning a collision repair?
   A. Determine the tools needed to make the repair.
   B. Make sure all the required materials are on hand.
   C. Determine how long the repair will take.
   D. Take work breaks between every repair procedure/step.

_____ 10. *True or False?* Body side molding can be bolted, clipped, or glued in place.

_____ 11. When trying to access damage on a vehicle door panel, which of the following pieces will *not* need to be removed?
   A. Window crank handle.
   B. Rear window.
   C. Door trim panel.
   D. Power window/lock wiring harness.

12. The vehicle panel repair process can be divided into what two steps?

_____

_____ 13. A few operations that may be involved in the _____ phase of vehicle panel repair include pulling, leveling, and edge alignment.
   A. painting
   B. finishing
   C. disposal
   D. roughing out

_____ 14. A corrective pull will always try to form a _____ line between the pull location and the anchor location on the vehicle.
   A. straight
   B. wavy
   C. curved
   D. None of the above.

_____ 15. *True or False?* The lift reaction has the greatest effect when there is a large angle between the attachment and the pull.

16. Why should a technician use a large contact area to pull?

_____

17. What are the two types of corrective force used to pull an attachment in a collision repair shop?

_____

_____ 18. *True or False?* A force will always move the strongest object first.

_____ 19. During a tension pull, a technician should insert _____ under the suspension of the vehicle to keep the vehicle at the correct height and prevent suspension sag.

20. What can happen if only a corner of a duckbill movable jaw contacts the panel area to be pushed?

_____

Chapter 9 Nonstructural Panel Repair　　　　　　　　　　　　　　　　　　　　　　65

Name _____

21. The repair process stretching is used to repair what type of metal damage?
_____

22. What is a *false stretch*, or *"oil can"*?
_____
_____

23. List three repair procedures to remove false stretch.
_____
_____
_____

24. What does the dolly do during leveling?
_____
_____

25. What does the hammer do during leveling?
_____
_____

_____ 26. *True or False?* Restore damaged crown by raising a single point in the center.

_____ 27. Repairing a hole caused by pulling on a draw pin can be done with a MIG welder or waterproof _____.

_____ 28. *True or False?* Arrowheads in the metal above and below a damaged body line can be released once the body line has been roughed out.

_____ 29. The act of reducing the surface area of metal during a collision repair is known as _____.

30. What is *warp damage*?
_____
_____

_____ 31. Which of the following tools can be used to heat shrink a stretched metal area?
　　A. Draw pin welder.
　　B. Micro torch.
　　C. Oxyacetylene torch.
　　D. All of the above.

_____ 32. *True or False?* As steel heats up, its color changes from silver to green and then to purple.

_____ 33. While shrinking, using a wet rag to cool the area while the metal is too hot with color can make the metal _____.

_____ 34. *True or False?* Shrinking metal with heat will *not* burn rustproofing from the back side of a panel.

35. Explain hail dent removal from a low or medium crown panel with a butane micro torch.
_____
_____
_____

36. What is the purpose of the dolly when using a pick hammer to shrink?
_____
_____

37. Describe *chasing the dent*.
_____
_____
_____
_____
_____

38. Explain how to prevent *chasing the dent*.
_____
_____
_____
_____

_____ 39. Excessive sanding of body filler leads to _____, a condition in which filler is sanded below the level of the adjacent metal.

_____ 40. _____ raises small lows and lowers small highs at the end of the roughing out process.

_____ 41. _____ is used to identify high areas and low areas at the end of the roughing out process.

42. What is *blind picking*?
_____
_____
_____

_____ 43. Which collision repair tool offers a very accurate alternative to blind picking when the back of the vehicle panel can be easily accessed?
　　A. Air saw.
　　B. Grinder.
　　C. Bull's-eye pick.
　　D. Electric shears.

Chapter 9 Nonstructural Panel Repair   67

Name _____

For questions 44–48, match the following "roughing out" processes with their descriptions.

_____ 44. Allows vehicle panels to fit and function properly if sides are damaged.

_____ 45. Relies on the lift reaction to restore panel dimension.

_____ 46. Uses heat to deal with false stretch, but too much can cause warp damage.

_____ 47. Lifts low areas to match surrounding areas.

_____ 48. Detailed last "roughing out" process to ensure a smooth panel finish.

A. Raising
B. Picking and filing
C. Shrinking
D. Edge alignment
E. Pulling

_____ 49. Which type of vehicle panel requires a one-sided repair?
A. Boxed.
B. Quarter panel behind the rear wheel.
C. Hood.
D. None of the above.

_____ 50. If an internal brace or frame rail is damaged, it must be straightened _____ the vehicle panel is roughed out.

51. What is the purpose of roughing?

_____
_____

52. When leveling in a one-sided repair of a dent, what takes the place of the dolly?

_____
_____

_____ 53. *True or False?* When fixing a dent with a one-sided repair that involves a body line, the body line should be raised last.

_____ 54. When raising a dent during a two-sided repair without a body line, use a dolly that is large enough to span the _____.
A. dent
B. hammer face
C. whole panel
D. center of the panel

_____ 55. The strongest point of arrowhead damage is area at the _____ of the arrowhead.

56. List the basic steps in a two-sided arrowhead repair that does not involve a body line.

_____
_____

Copyright by Goodheart-Willcox Co., Inc. May not be reproduced or posted to a publicly accessible website.

_____ 57. *True or False?* When a technician makes a repair to a fold in a panel, he/she should use corrective force that is the opposite of the force that created the damage.

_____ 58. When repairing a fold on a vehicle panel with an internal brace, the _____ is the strongest part of the damage.

59. List the two ways to finish a vehicle panel.
_____

60. When metal finishing, explain how to remove high or low areas that are larger than 1/8″ in diameter.
_____
_____

_____ 61. When filling, body filler should not be thicker than _____ inch.

_____ 62. Scratches in the bare metal anchor the body filler by _____ adhesion.

63. What is the bluish-colored liquid in a can of filler?
_____

_____ 64. *True or False?* The curing process occurs one hour after the hardener and filler are mixed.

_____ 65. After applying the first thin coat of filler to an area, additional coats should be applied, building up the filler until a(n) _____ is achieved.
A. crater
B. overfill
C. underfill
D. dip

66. What should a technician do if the filler feels soft, moist or sticky after scratching it with a fingernail?
_____

_____ 67. _____ sanding levels filler and can be done with hand or power tools.

68. Explain how a conformable sander can form body filler to match a vehicle body line.
_____
_____
_____
_____
_____
_____
_____

Chapter 9 Nonstructural Panel Repair

Name _____

_____ 69. _____ are air pockets or voids in cured body filler.

_____ 70. When applying lead filler, the key is learning to _____ used to soften the lead stick.
A. control the cold
B. minimize the pressure
C. maximize the pressure
D. control the heat

_____ 71. When melting lead, work in a well-ventilated area and wear approved _____ protective equipment to prevent inhalation of lead dust or vapors.

_____ 72. All of the following are steps in lead filler application *except*:
A. smoothing.
B. tinning.
C. cracking.
D. stripping.

_____ 73. *True or False?* Aluminum is lighter and easier to recycle than steel.

_____ 74. *True or False?* Use the same body file on steel and aluminum.

_____ 75. An aluminum vehicle panel should be coated with a(n) _____ before applying body filler.

_____ 76. Do not heat an aluminum panel to more than _____ °F.

_____ 77. *True or False?* The infrared thermometer should be pointed at bare metal to get an accurate reading while heating an aluminum panel.

_____ 78. When repairing an aluminum edge with a heat gun, start heating at the _____ of the bend and work toward the center.

79. What should be checked before using a heat gun to repair hail damage on an aluminum panel?
_____

_____ 80. During dent repair on an aluminum panel, what is used to determine if a low area has been sufficiently raised?
A. Straightedge.
B. Dolly.
C. Level.
D. Rule.

Name _____ Date _____ Class _____

Chapter 10

Bolted Nonstructural Part Replacement

Learning Objectives
After studying this chapter, you will be able to:
- Identify part sources.
- Apply basic skills used in bolted part replacement.
- Identify parts of bolted panel assemblies.
- Explain how to remove damaged bolted parts.
- Explain how to install and align bolted parts.

Carefully read Chapter 10 of the textbook and then answer the following questions in the space provided.

_____ 1. Since replacement parts must be obtained within a reasonable amount of time, _____ is a key issue when deciding on a part source.

2. What do the letters OEM stand for?
_____

_____ 3. Most collision repair shops cultivate a relationship with _____ because the shop relies heavily on this source's ability to supply required OEM parts in a timely manner.

_____ 4. Used parts are sold as part of a(n) _____, not as individual pieces.

_____ 5. *True or False?* Used parts may have previously repaired damage and should be checked carefully before installation.

6. Define the term *aftermarket parts*.
_____
_____
_____

_____ 7. Vehicle parts from which of the following sources are generally the most expensive?
　　A. Original equipment manufacturer.
　　B. Used.
　　C. Aftermarket.
　　D. Salvaged.

8. What is the first step in replacing a bolted nonstructural part?
   _____
   _____
   _____

9. What should you look for when analyzing damage?
   _____
   _____
   _____

_____ 10. *True or False?* When removing bulbs to repair damage on a vehicle, take the bulb out of its housing and allow it to hang freely.

_____ 11. *True or False?* A bolt-on part will fit properly only if the parts it attaches to are in proper alignment.

_____ 12. All panel gaps should be _____ in width on a properly aligned repair.
   A. shifting
   B. varied
   C. tapered
   D. uniform

_____ 13. *True or False?* The concept of level in autobody uses the ground as a reference.

14. Explain how to sight a bumper for level.
   _____
   _____
   _____
   _____
   _____

_____ 15. _____ means that two or more adjacent panels are at the same surface level.

16. What can be caused by a vehicle door positioned farther out than the fender?
   _____

_____ 17. A front bumper is also called a(n) _____.

_____ 18. *True or False?* If a force impacts a non-isolator steel bumper that is greater than the bumper's yield point, the bumper will be damaged.

19. List two basic goals in removing any damaged vehicle part.
   _____
   _____

Chapter 10 Bolted Nonstructural Part Replacement

Name _____

_____ 20. When installing a steel bumper on the front of a vehicle, the _____ and the bumper end need to align.
   A. windshield
   B. deck lid
   C. wheel openings
   D. All of the above.

21. What is a *J-clip/nut*?
   _____
   _____
   _____

22. List the three parts of a flexible bumper assembly.
   _____
   _____
   _____

23. What is the first step in removing a plastic bumper?
   _____
   _____

_____ 24. When removing bumper cover retainers, a(n) _____ can be used to raise the center stud of the retainer.
   A. air saw
   B. upholstery tool
   C. body hammer
   D. tap and die set

_____ 25. The _____ is also called the headlight mounting panel or the grille panel.

_____ 26. *True or False?* If a fender and the header panel are both damaged, remove the fender and header together.

_____ 27. Header panel alignment is dependent on the position of the fenders and _____.

28. What is the difference between *skin* and *reinforced* steel fenders?
   _____
   _____
   _____
   _____

29. What may hold a fender liner in place?
   _____

_____ 30. *True or False?* Removing the parts on the original fender and placing them on the shop floor prevents them from being lost.

31. List the first gap to check in fender alignment.
_____

_____ 32. *True or False?* If a fender is not flush with the hood, then the fender is usually adjusted to fit the hood.

33. Explain how to adjust a fender edge if the fender edge is positioned inward compared to the vehicle door.
_____
_____
_____
_____
_____
_____

_____ 34. A vehicle hood is held shut by a(n) _____ and a hood latch.

_____ 35. If a damaged hood will not open, which of the following should a technician try?
   A. Pull on the hood release while another pulls up on the hood.
   B. Remove the cable release and unbolt the hood latch.
   C. Insert a long pry bar under the hood and pry open the hood.
   D. All of the above.

_____ 36. *True or False?* Remove the hood latch before installing the hood.

37. Why should a technician make an X-check of the engine compartment before installing a hood?
_____
_____
_____

For questions 38–41, match the following hood alignment checks with their sequence.

_____ 38. Front gap.

_____ 39. Fender gaps.

_____ 40. Hood flush with fenders.

_____ 41. Cowl vent panel or windshield gap.

A. First check
B. Second check
C. Third check
D. Last check

_____ 42. A vehicle hood should latch completely when it is dropped from about _____ above the radiator support.

Chapter 10 Bolted Nonstructural Part Replacement

Name _____

43. What happens to the hood height at the front if a technician places a shim between the hood and the rear hood bolt?

_____

_____ 44. The _____ is the framework that supports all other door components.

_____ 45. In the event of a side impact collision, a _____ within the vehicle door helps protect passengers from injury.
   A. crash beam
   B. door skin
   C. window frame
   D. belt molding

46. Explain how to check a door hinge for wear.

_____

_____

_____

_____ 47. *True or False?* The first step in removing a rivet is to drill out the center hole.

48. List the alignment checks to make upon installing a door.

_____

_____

_____

_____ 49. When checking how a door seals, close the door on a sheet of paper and then pull the paper out. There should be a(n) _____ drag.
   A. slight
   B. strong
   C. nonexistent
   D. overpowering

_____ 50. Adding a shim between the top hinge and the vehicle door moves the rear of the door _____.
   A. upward
   B. downward
   C. inward
   D. outward

51. No alignment bar is available to raise the rear of a door with welded-on hinges. What tool can a technician use?

_____

_____ 52. A vehicle door should shut when it is pushed from a distance of about _____ feet.

Copyright by Goodheart-Willcox Co., Inc. May not be reproduced or posted to a publicly accessible website.

_____ 53. If there is difficulty opening or closing a door, moving the striker up will bring the door _____.
   A. upward
   B. outward
   C. downward
   D. inward

54. What should you check if a power window will not go all the way down after a door replacement?
   _____

55. List the alignment checks to make after replacing a deck lid.
   _____
   _____
   _____
   _____

56. Explain how to find the source of a deck lid water leak.
   _____
   _____
   _____
   _____

_____ 57. *True or False?* A new rear hatch should be painted underneath and around the window opening before installation.

_____ 58. A sliding vehicle door moves on upper and _____ rollers.
   A. front
   B. lower
   C. rear
   D. None of the above.

59. List the parts that must be removed before removing a pickup bed.
   _____
   _____
   _____
   _____

_____ 60. *True or False?* After a new pickup bed is installed, the gap between the cab and the bed should taper to one side when aligned correctly.

Name _____ Date _____ Class _____

Chapter 11

Welded and Bonded Nonstructural Panel Replacement

Learning Objectives

After studying this chapter, you will be able to:
- List the steps involved in replacing welded and bonded nonstructural panels.
- Analyze damage that requires panel replacement.
- Locate and remove spot welds.
- Describe various panel joining techniques.
- Explain the procedures for replacing specific welded and bonded nonstructural panels.

Carefully read Chapter 11 of the textbook and then answer the following questions in the space provided.

_____ 1. In a(n) _____, the entire damaged panel is removed from the vehicle at the factory seams and a replacement panel is welded or bonded in place.

_____ 2. In a(n) _____, the damaged portion of a panel is removed and only a portion of the full replacement panel is installed.

3. List the general steps to replace damaged welded or bonded structural panels.

_____
_____
_____
_____

_____ 4. *True or False?* Since most vehicle panels are connected, damage to one panel stays confined to that specific panel.

_____ 5. The area where the vehicle and the replacement part meet is called the _____.

6. Identify the joint types shown in the following illustrations.

   A. _____
   B. _____
   C. _____
   D. _____
   E. _____

_____ 7. Which of the following joints has a step?
   A. Butt joint.
   B. Sleeve joint.
   C. Flange joint.
   D. None of the above.

8. What is a *sleeve*?

   _____

_____ 9. *True or False?* In a lap joint, the top panel is flush with the bottom panel.

_____ 10. The type of joint made will dictate how the panels are _____.

11. If there is damage that extends into two adjacent panels that are welded together, what should be done before removing the damaged panel?

   _____

   _____

_____ 12. *True or False?* If a full panel replacement is made, rough cutting is not normally required because the joint locations are already determined.

_____ 13. If the replacement panel is new, rough cutting involves cutting the panel approximately 1″–2″ _____ than required.

14. Why should a technician write on the masking tape used to indicate a cut line?

   _____

   _____

   _____

Chapter 11 Welded and Bonded Nonstructural Panel Replacement

Name _____

15. Briefly explain how a factory spot weld is made.

_____
_____
_____
_____
_____
_____

16. How can a technician locate spot welds?

_____
_____
_____
_____

_____ 17. *True or False?* During welded part replacement, the underlying panel should, in most cases, be preserved as well as possible.

_____ 18. During removal of spot welds, a(n) _____ drilled in the weld will prevent a hole saw from walking off the weld.

_____ 19. After cutting, a(n) _____ will leave a nugget on the underlying panel once the upper panel is removed.

_____ 20. *True or False?* When chiseling a spot weld, the air chisel must be kept level.

_____ 21. *True or False?* When replacement panel test fitting, there should be a gap between the mounting flanges and the panel.

22. In final cutting, how much overlap is required for a lap joint?

_____

_____ 23. *True or False?* Use a grinder to remove paint from a flange joint.

24. Why should a technician always hold the flange tool against the edge of the panels?

_____
_____
_____

25. What does a flange tool do to a high-crown panel?

_____
_____

_____ 26. Which type of panel corner allows a flange tool to make a continuous bend?
A. Square.
B. Rounded.
C. Jagged.
D. None of the above.

27. Why is weld-through primer sprayed on bare metal vehicle panels?
_____
_____
_____

_____ 28. *True or False?* Before welding, scratch the weld-through primer to allow a MIG welder to strike an arc.

29. How does a technician prevent butt-welded panels from rising or lowering from the heat of welding?
_____
_____
_____

30. List two tools that hold replacement parts in position when welding.
_____
_____

31. Why must a technician check the alignment of every part before welding?
_____
_____

_____ 32. _____ involves the use of adhesives to join replacement parts to a vehicle.

_____ 33. The combination of squeeze-type resistance spot welding and the use of adhesives to join replacement parts to a vehicle is known as _____.

_____ 34. The strength of the _____ determines the quality of the replacement panel repair and many other metal collision repairs.

35. If MIG plug welds will be used to replace factory spot welds in a repair, how many welds should be used and where should they be placed?
_____
_____
_____

_____ 36. *True or False?* When making a bonding repair, the first two inches of mixed adhesive squeezed from an adhesive gun should be discarded.

Chapter 11 Welded and Bonded Nonstructural Panel Replacement

Name _____

37. After adhesive has contacted both panels to be bonded, how can a technician make a slight adjustment?

_____

_____

_____

38. How much of an overlap is needed for a bonded lap joint?

_____

39. What does a featheredge do for repairing screw holes left from holding panels in position during installation?

_____

_____

_____

_____ 40. In the process of weld bonding, if the spacing of the welds is not specified by the manufacturer, place one weld every _____ inches.

_____ 41. Dressing welds involves grinding welds and removing _____.

42. On a cutoff tool, why is a grinding wheel preferred to a grinding disk to finish MIG welds?

_____

_____

_____ 43. *True or False?* Welding will burn the paint protection off the back side of a vehicle panel.

_____ 44. Which of the following items may be found inside a vehicle quarter panel during quarter panel replacement?
   A. Trunk liner.
   B. Trunk weather strip.
   C. Door striker.
   D. All of the above.

_____ 45. *True or False?* Quarter panels are mainly installed as a full panel replacement.

_____ 46. Due to its location on the vehicle, damage to the quarter panel will often extend into the _____.
   A. front bumper
   B. grille panel
   C. wheel house
   D. windshield

_____ 47. The gap between the panels in a sleeve joint should be _____″ for the weld.

_____ 48. The bows and headers that strengthen a vehicle roof are attached to the roof panel with a soft adhesive that acts as a(n) _____ material.

_____ 49. Which of the following types is *not* a way that a roof can be joined to a vehicle?
   A. Lap joint.
   B. Gutter joint.
   C. Tracker joint.
   D. Wraparound joint.

50. What type of roof is found on a uniside vehicle?
_____

51. Explain the two types of flanges that may be found between a roof and a sail panel.
_____
_____
_____

52. What items are required to be removed as part of the "quick method" of roof replacement?
_____
_____
_____
_____

_____ 53. *True or False?* A MIG weld will not stick to the silicon bronze applied at the joint area of a windshield pillar and a sail panel.

_____ 54. A vehicle rear body panel is *not* spot welded to which of the following parts?
   A. Front door panel.
   B. Quarter panels.
   C. End of the rear frame rails.
   D. Trunk floor.

_____ 55. *True or False?* Check alignment and the deck lid, taillights, and rear bumper for fit before welding a replacement rear body panel.

_____ 56. *True or False?* If a replacement bed skin is difficult to install, it should be bonded rather than welded.

_____ 57. The cab corner on a pickup truck is similar to the _____ on a car.
   A. deck lid
   B. hood
   C. roof
   D. quarter panel

_____ 58. Some technicians use a(n) _____ to hide a splice joint in a cab corner replacement.

Chapter 11 Welded and Bonded Nonstructural Panel Replacement 83

Name _____

59. List what checks should be made to the replacement cab corner before welding.

_____

_____

_____

_____ 60. Deciding if a(n) _____ can be made in the window frame during a door skin replacement will eliminate the need to blend paint into the roof.
   A. weld
   B. splice
   C. crease
   D. bend

61. When skinning a door, why should a technician leave the paint edge in place?

_____

_____

_____ 62. The _____ is the number of minutes between the application of an adhesive and the time the adhesive begins to cure.

_____ 63. *True or False?* Once an adhesive makes contact with a door skin, reposition the skin only by lifting.

_____ 64. Before making a patch panel repair, all _____ metal must be identified and cut out.
   A. rust-weakened
   B. thick
   C. painted
   D. bent

Name _____ Date _____ Class _____

Chapter 12

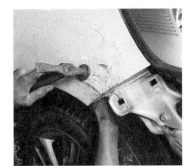

Plastic Repair

Learning Objectives

After studying this chapter, you will be able to:
- Describe the various ways to classify plastics.
- Recognize tools and materials used in plastic repair.
- Summarize the basic steps used in plastic repair.
- Describe procedures for repairing specific types of plastic damage.

Carefully read Chapter 12 of the textbook and then answer the following questions in the space provided.

_____ 1. *True or False?* Plastic parts are less susceptible to minor dents than metal parts.

2. List the three causes of poor adhesion.

_____

_____

_____ 3. Which of the following can be bent to 180° without breaking?
   A. Rigid plastic.
   B. Semirigid plastic.
   C. Flexible plastic.
   D. All of the above.

_____ 4. Which of the following about thermoplastics is true?
   A. They can be heated to melting temperatures.
   B. They can retain all their original characteristics after heating.
   C. They can be welded with special plastic welding equipment.
   D. All of the above.

_____ 5. *True or False?* Thermoset plastics burn rather than melt when heated.

_____ 6. *True or False?* All thermoset plastics are rigid.

Copyright by Goodheart-Willcox Co., Inc. May not be reproduced or posted to a publicly accessible website.   85

For questions 7–11, match the automotive plastic with its part type.

_____ 7. Fender liners, radiator shrouds.

_____ 8. Running boards.

_____ 9. Hood, header, hatch, fender.

_____ 10. Grilles, interior trim.

_____ 11. Bumper cover, ground effects.

A. SMC (Sheet molded compound)
B. PUR (Polyurethane)
C. ABS (Acrylonitrile butadiene styrene)
D. UP (Fiberglass)
E. PP (Polyproylene)

_____ 12. If a bumper cover is yellow on the back side, it is made of _____.
   A. urethane
   B. TEO
   C. TPO
   D. xenoy

_____ 13. Stretch damage occurs in which type of plastic?
   A. SMC.
   B. ABS.
   C. Fiberglass.
   D. Polyurethane.

_____ 14. Thermo elastomeric olefin (TEO) is used for _____.

_____ 15. Overflow tanks and fender liners are made from _____ plastic.

16. Identify the parts of the hot-air welder in the following illustration.
   A. _____
   B. _____
   C. _____
   D. _____
   E. _____

17. What repair materials are used for SMC plastic?

_____

18. What repair materials are used for ABS plastic?

_____

19. Which welding shoe is used to melt and shape the filler rod?

_____

Chapter 12 Plastic Repair

Name _____

_____ 20. *True or False?* You must identify plastic before welding with universal rod.

21. Which type of welder heats a larger area than an airless welder and requires more skill to operate?

_____

_____ 22. The adhesive has a(n) _____, which is the number of minutes between mixing and curing.

_____ 23. Adhesion promoter is used _____ applying adhesives to olefin-containing plastics.

24. Which mesh can be welded into plastic?

_____

_____ 25. *True or False?* The fiberglass cloth or mat acts as a reinforcement for the resin and hardener mixture.

26. What is the first step in all plastic repair?

_____

27. Explain two ways to remove a bumper cover.

_____

_____

_____

_____

28. Explain how to use a grinder to check for olefin in a part.

_____

_____

_____

_____ 29. *True or False?* If the plastic floats, it contains olefin.

_____ 30. Damaged plastic must be _____ before adhesive or welding repairs can be made.

31. If aluminum tape is not strong enough to hold damaged plastic in place, what can be used?

_____

_____ 32. The most common plastic alignment problem is a difference in _____.

_____ 33. *True or False?* Flexible plastic will occasionally spring back into position once a force is removed.

Copyright by Goodheart-Willcox Co., Inc. May not be reproduced or posted to a publicly accessible website.

34. What is passive shaping?

_____

_____

35. How do you determine if a heated plastic part is hot enough for shaping?

_____

_____

_____ 36. A(n) _____ is a piece of plastic that is the same type of material as the damaged part.

37. What is used to increase the contact surface area of the adhesive or weld repair?

_____

_____ 38. *True or False?* V-grooves are made on the front and back of a repair area.

_____ 39. _____ produces a gradual slope.
   A. Welding
   B. Binding
   C. Beveling
   D. Bonding

_____ 40. Which repair method only melts the filler rod?
   A. V-groove.
   B. Welding.
   C. Beveling.
   D. Binding.

_____ 41. Which repair method melts both the filler rod and base material?
   A. Binding.
   B. Welding.
   C. Beveling.
   D. Bonding.

_____ 42. *True or False?* Weldable plastic cannot be welded to itself.

_____ 43. *True or False?* The welding rod should change color as it is melted in the shoe.

44. What type of weld is used on ABS plastic only?

_____

_____ 45. *True or False?* There are four components to a good weld: temperature, angle, pressure, and speed.

46. If a plastic part being bonded contains olefin, what must be applied to the repair area?

_____

_____ 47. _____ is the process of making a part.

Chapter 12 Plastic Repair

Name _____

48. What is the last step in plastic repair?

_____

49. What type of sanding should be used on olefin plastic?

_____

_____ 50. *True or False?* Lowering air pressure, decreasing gun speed, decreasing distance, and increasing the number of coats decreases the grain size of the textured plastic.

51. How do you clean an olefin-containing part with gouge damage?

_____
_____
_____

_____ 52. *True or False?* Some adhesive manufacturers recommend that the adhesive be applied to the paint along the featheredge.

_____ 53. *True or False?* To use a universal rod to repair the tear in the TPO bumper cover, cut a V-groove in the front side of the break.

54. What should you do if the edges of a binding repair will not featheredge?

_____
_____

_____ 55. *True or False?* Adhesives can be used to repair a tear in flexible plastics only.

_____ 56. *True or False?* Sanding with a DA may cause plastic to melt.

_____ 57. When a bumper cover moves during a collision, the mounting _____ of the bumper cover does not move and may be torn from the bumper cover.

_____ 58. ABS grilles are often coated with _____ finish.

_____ 59. *True or False?* Most body filler will adhere to FRP.

_____ 60. *True or False?* Most body filler will adhere to SMC.

61. A plastic part has inaccessible internal damage. What should you do?

_____

62. When repairing rigid SMC parts, what should you do if unbroken fibers prevent alignment?

_____
_____

_____ 63. When the damage to a SMC part is so great that the damaged area must be cut out and replaced, the process is known as _____.

Name _____ Date _____ Class _____

Chapter 13

Glass

Learning Objectives

After studying this chapter, you will be able to:
- Describe the types of glass found on vehicles.
- Identify the specialized tools and materials used in glass service.
- Demonstrate the procedure for removing and replacing movable glass.
- Explain how to remove and replace fixed glass.

Carefully read Chapter 13 of the textbook and then answer the following questions in the space provided.

_____ 1. *True or False?* On unibody vehicles, the glass may add to the structural strength of the unibody.

_____ 2. _____ consists of two layers of glass with a layer of polyvinyl butyl plastic between them.

_____ 3. What serves as a backstop to support an inflated passenger side air bag?
A. Dashboard.
B. Windshield.
C. Neither A nor B.
D. Both A and B.

4. How is safety glass tempered?
_____

5. What does the chemical film on coated glass do?
_____

_____ 6. *True or False?* Encapsulated glass must be removed and installed as a unit.

7. Name the two categories of automotive glass.
_____

_____ 8. *True or False?* Encapsulated glass is movable glass.

_____ 9. *True or False?* Moving glass regulators can be operated manually only.

_____ 10. Rolling glass moves up and down in _____.

_____ 11. *True or False?* Weather strip seals the movable glass to prevent water leaks, but not air leaks.

_____ 12. Which of the following is a type of fixed glass?
   A. Bolted glass.
   B. Rubber-set glass.
   C. Urethane-set glass.
   D. All of the above.

_____ 13. *True or False?* Butyl tape is stronger than urethane.

_____ 14. *True or False?* An improperly installed urethane-set windshield can be pushed loose from the pinch weld by the force of a passenger-side air bag deployment.

_____ 15. *True or False?* Urethane is strengthened by sunlight.

_____ 16. A(n) _____ is a black border around the perimeter of any urethane-set glass.

_____ 17. *True or False?* The glass shown in the following illustration has no molding.

_____ 18. A failure in the following may cause urethane-set glass to fail during a collision *except*:
   A. pinch welds.
   B. weather strip seals.
   C. glass primer.
   D. urethane.

_____ 19. *True or False?* A plastic paddle can scratch surfaces like the hook tool.

_____ 20. *True or False?* A windshield is too strong to be broken by a cut-out knife.

_____ 21. When using a banana knife, lubrication with _____ will make cutting easier.

Chapter 13 Glass 93

Name _____

_____ 22. *True or False?* When using a windshield wire tool, the urethane is cut with a sawing action.

_____ 23. A(n) _____ can be used to affix a handle on glass.

24. Identify the tool shown in the following illustration.

_____

_____ 25. Pulling the caulking gun's trigger ratchets the plunger into the _____ tube.

_____ 26. An oscillating knife has a blade that moves from _____.
  A. side to side
  B. front to back
  C. top to bottom
  D. None of the above.

_____ 27. *True or False?* Use only one manufacturer's glass repair products on any given job.

_____ 28. All of the following are used as lubricants to help slip glass inside a gasket *except*:
  A. white grease.
  B. soap and water.
  C. glass cleaner.
  D. silicon spray.

_____ 29. The amount of time required for the solvents to evaporate from a paint or primer is known as _____.

_____ 30. When repairing glass, _____ must be applied to any bare metal on the pinch weld before the glass is installed.

_____ 31. *True or False?* Damming tape is needed when using high-viscosity urethane.

_____ 32. The lap shear strength of cured urethane is _____ psi.

_____ 33. *True or False?* The customer should not drive the vehicle until the urethane on the repaired windshield has cured.

_____ 34. To remove rolling glass in a door, roll down the window and remove the _____.

_____ 35. *True or False?* When drilling out rivets, the drill bit should be slightly smaller than the diameter of the rivets' center hole.

_____ 36. A(n) _____ is a plastic clip that snaps around the edge of the door glass.

_____ 37. Sliding glass may be _____.
　　A. bolted in place
　　B. set in rubber
　　C. Neither A nor B.
　　D. Both A and B.

_____ 38. *True or False?* Hinged glass screws are made of plastic and are easily stripped.

_____ 39. Turning glass is secured by a(n) _____.

_____ 40. *True or False?* When removing rubber-set glass, push the corner of the gasket and the glass toward the inside of the vehicle.

_____ 41. *True or False?* When installing rubber-set glass, insert a piece of rope into the pinch weld channel of the gasket.

_____ 42. *True or False?* When removing rubber-set glass, work from inside the vehicle and start at the corner with the smallest radius, peel the gasket back and push the glass in.

43. Name the two methods used to replace urethane-set glass.

_____

_____ 44. Some windshields have _____ blocks, which are placed at the bottom of the glass.

_____ 45. When replacing urethane-set glass, apply the urethane to the pinch weld, and cut the nozzle in the shape of a(n) _____.
　　A. "C"
　　B. "U"
　　C. "D"
　　D. "V"

_____ 46. When applying urethane to a windshield or back glass opening, start at the _____.
　　A. top center
　　B. bottom center
　　C. passenger side center
　　D. driver side center

_____ 47. When cutting out encapsulated glass, work from _____ the vehicle.

48. When can leak checks be performed on new installations of encapsulated glass?

_____

_____

Chapter 13 Glass

Name _____

_____ 49. All of the following are methods of leak detection *except*:
   A. audio.
   B. air.
   C. ultrasonic.
   D. water.

_____ 50. *True or False?* An ultrasonic signal generator locates leaks by sending out inaudible sound waves.

For questions 51–54, match the following types of glass with their descriptions.

_____ 51. Single layer of glass used for any window other than windshield.

_____ 52. Plastic frame molded around a piece of safety glass.

_____ 53. Two layers of glass with a layer of polyvinyl butyl plastic between them.

_____ 54. May contain an electrical grid.

A. Laminated safety glass
B. Tempered safety glass
C. Encapsulated glass
D. Heated glass

Name _____ Date _____ Class _____

Chapter 14

Unibody/Frame Straightening Equipment

Learning Objectives

After studying this chapter, you will be able to:
- Identify and explain how to use hook-up equipment.
- Identify and explain how to use tie-down equipment.
- List the types of pulls.
- Identify various types of frame machines.

Carefully read Chapter 14 of the textbook and then answer the following questions in the space provided.

_____ 1. The capacity of most rams used in structural repair is _____ tons.

_____ 2. *True or False?* All equipment used in structural repair must be able to withstand 20 tons of pulling force.

_____ 3. Pulling chains are made from heat-treated alloy steel and have a minimum tensile strength of _____ lbs.
   A. 5,000
   B. 10,000
   C. 25,000
   D. 50,000

_____ 4. The safety chain prevents _____ if the chain breaks or the hookup fails.

_____ 5. A(n) _____ chain is a supplemental chain or cable that is connected to the pull chain.

_____ 6. *True or False?* A chain should never be heated.

_____ 7. Never stand between the point where the pulling chain is hooked up to the vehicle and the _____.

_____ 8. The _____ hook is used to attach a chain to a clamp or a chain loop.

9. What type of hook attaches chains to other chains?

_____

_____ 10. The _____ hook can be used to tie two chains together or to shorten a length of chain.

_____ 11. _____ are used to hold the damaged vehicle in place so that pulls can be made to align the damage.

_____ 12. In most cases, tie-downs are made in the _____ section of the vehicle.
   A. front
   B. back
   C. center
   D. Either A or B.

_____ 13. _____ can be used to tie-down a full-frame vehicle.

_____ 14. Placing _____ under the frame prevents downward movement of the frame at its location.

_____ 15. *True or False?* During blocking, the suspension is stressed because it does not move down.

16. How do you prevent damage to frame flanges when using a frame wrap?

_____
_____
_____

_____ 17. What size chain should be used to make tie-downs?
   A. 3/16".
   B. 1/8".
   C. 1/9".
   D. 3/8".

18. Where do you place a tie-down hook on a full-frame?

_____
_____
_____

_____ 19. Most unibody vehicles are tied down with clamps that connect to the _____ in the rocker panels.

_____ 20. Tie-down clamps are placed at the _____ of the vehicle's center section.
   A. front
   B. back
   C. corners
   D. Both A and B.

_____ 21. What grade of bolt do you use to replace a clamp bolt?
   A. 2.
   B. 4.
   C. 6.
   D. 8.

Chapter 14 Unibody/Frame Straightening Equipment

Name _____

_____ 22. _____ are used to attach the pull to the damage.

_____ 23. *True or False?* Hookups should be made at the weakest area available.

_____ 24. Which clamp requires a hole to be drilled in the unirail?
A. Self-tightening.
B. Sandwich.
C. Rotating pull.
D. Wedge.

25. How does a self-tightening clamp stay in place on a unirail?

_____

_____

_____ 26. *True or False?* A pull plate allows for few bolt combinations.

_____ 27. Slings can be made from _____.
A. steel cable
B. nylon
C. plastic
D. Both A and B.

_____ 28. *True or False?* A sling will loosen as it is pulled.

29. How do you make a frame wrap?

_____

_____

_____

_____ 30. A(n) _____ pull involves wrapping a chain around the frame and tying the chain to itself.

_____ 31. *True or False?* The four-ton hydraulic rams used in nonstructural repair are powered by air-powered pumps.

_____ 32. A short loop on a frame wrap may cause a side pull to _____.

33. What does a gauge on a frame machine indicate?

_____

_____

_____ 34. In a full-frame repair, the ram is set _____ the frame rails.
A. on top of
B. in between
C. underneath
D. across

35. When using a ram, how can you ensure that the damaged area moves instead of the base?
    A. Use a tie-down to immobilize the undamaged frame rail.
    B. Weaken the damage area by heating.
    C. Both A and B.
    D. Neither A nor B.

_____ 36. *True or False?* The action of the pull ram is similar to that of a conventional ram.

_____ 37. *True or False?* A pull ram attached to the floor can make a downward pull on a frame rail that has been pushed up by collision damage.

_____ 38. The power behind the straightening process is known as _____.

39. What does the phrase "run out of ram" mean?

_____

_____

_____ 40. On a(n) _____ pull, the chain is stretched from the hookup to a chain anchor.

_____ 41. *True or False?* In a vector pull, a higher pull has more power.

_____ 42. A frame machine that can be moved from one stall to another is known as a(n) _____ system.

_____ 43. What pull is not possible with a sliding-arm portable system?
    A. Forward.
    B. Backward.
    C. Upward.
    D. Downward.

44. Describe a floor pot system.

_____

_____

45. Describe a frame bench.

_____

_____

_____ 46. Which bench uses specific fixtures?
    A. Universal.
    B. Drive on.
    C. Dedicated.
    D. Conventional.

_____ 47. *True or False?* A universal bench has fixtures.

_____ 48. A(n) _____ rack is a holding and pulling machine that the vehicle can be driven onto or pulled onto with a winch.

Chapter 14 Unibody/Frame Straightening Equipment 101

Name _____

_____ 49. *True or False?* All frame racks have 360° pulling capabilities.

_____ 50. In a(n) _____ rail system, the vehicle is positioned within the rail grid.

51. Identify the tie-downs, clamps, and hooks shown in the following illustrations.

A. _____
B. _____
C. _____
D. _____
E. _____

Name _____ Date _____ Class _____

Chapter 15

Measurements

Learning Objectives

After studying this chapter, you will be able to:
- Understand measurements listed in a dimension manual.
- Make linear measurements and compare them to a standard.
- Install and interpret frame gauges.
- Set up and read a mechanical 3-D measuring system.
- Install and use a computer-aided measuring system.
- Draw a damage diagram.

Carefully read Chapter 15 of the textbook and then answer the following questions in the space provided.

_____ 1. The _____ is an imaginary plane that runs down the center of a vehicle, dividing the vehicle into left and right halves.

_____ 2. *True or False?* The centerline is stamped or marked on the vehicle.

_____ 3. Width is measured _____ to centerline.

_____ 4. *True or False?* Height is the distance of a point below datum.

_____ 5. _____ is an imaginary plane underneath the vehicle.

_____ 6. *True or False?* Height measurements are taken parallel to centerline and perpendicular to datum.

_____ 7. _____ is the distance between two points measured parallel to the centerline.

_____ 8. The X-check will determine if the measured area is _____.

_____ 9. *True or False?* Symmetrical measuring points are the same distance from the centerline, but not the same height from datum.

_____ 10. *True or False?* Vehicles can be asymmetrical.

11. List the two sources for standards.
    _____

_____ 12. The linear measurements of width, height, and length are known as _____.

_____ 13. *True or False?* Underbody dimensions are taken from the engine compartment, door, hood, or deck lid openings.

_____ 14. *True or False?* One-dimensional measurements are measurements taken between two points.

_____ 15. *True or False?* When measuring between holes of the same size, measuring from the far edge of one to the near edge of the other is the same as measuring from center to center.

_____ 16. A(n) _____ is a measuring tool used to find the center of a round hole.

_____ 17. The maximum allowable deviation from the standard is known as _____.

_____ 18. The tolerance on a full-frame vehicle is _____″.
    A. 1/4
    B. 1/2
    C. 3/4
    D. 1

_____ 19. The tolerance on a unibody vehicle is _____ mm.
    A. 1
    B. 3
    C. 6
    D. 10

_____ 20. The tolerance for an X-check is _____ that for a linear measurement.

_____ 21. *True or False?* To make an X-check, use a tram with equal-length pointers.

_____ 22. Door-opening measurements are taken from _____.
    A. seams
    B. bolt holes
    C. striker
    D. All of the above.

23. Explain how to measure a windshield opening with round corners.
    _____
    _____
    _____
    _____

Chapter 15 Measurements

Name _____

_____ 24. *True or False?* The tape measure should be stretched tight and straight when measuring.

_____ 25. The _____ is a straight bar with a fixed pointer and a sliding pointer.

_____ 26. *True or False?* If the measuring points are at equal heights, the point-to-point measurements will be equal to the datum measurements.

27. What type of measurements use height and width dimensions to locate a measuring point?
_____

28. What does gauging measure?
_____
_____

29. List the parts of a centerline gauge.
_____
_____

30. Explain how to check a unibody vehicle for droop.
_____
_____
_____
_____
_____
_____

_____ 31. Gauges should be mounted on unibody vehicles with the vehicle _____.

_____ 32. When using gauges to check for damage, begin by performing an X-check under the _____ section of the vehicle.
A. front
B. rear
C. center
D. top

_____ 33. When gauging, if the center section is not _____, the front and rear sections will be misaligned.

_____ 34. If the legs of a full-frame vehicle center section X-check are off by more than _____ mm, diamond damage may be present.
A. 1
B. 2
C. 4
D. 6

_____ 35. The _____ centerline gauges determine if damage is present in the center section of the vehicle.

_____ 36. When the front and rear portions of the center section are not level in relation to each other, this is known as _____.

_____ 37. *True or False?* Diamond damage may cause the front and rear sections to rotate.

_____ 38. If a vehicle with out-of-level or twist damage shows centerline pin misalignment, the centerline pin may be off due to the _____ effect.

_____ 39. *True or False?* Mash damage has no effect on centerline pin alignment.

_____ 40. _____-dimensional measurements provide the most accurate damage information.
   A. One
   B. Two
   C. Three
   D. Four

41. What holds parts in the proper position so they can be welded together?

_____

_____ 42. A dedicated bench uses jig-like structures called _____ to identify damage and hold replacement parts in position.
   A. struts
   B. towers
   C. fixtures
   D. pins

_____ 43. The damaged vehicle is held in place for pulling by the _____ clamps, not by the fixtures.

_____ 44. A universal bench is similar to a dedicated bench in that the bed of the bench serves as a(n) _____ plane.

_____ 45. Holes located in the four corners of the vehicle's center section that will support the weight of the vehicle are called _____.

_____ 46. A(n) _____ measuring system relies on a beam of light to form a straight line from the emitter to the target.

47. What lists the location of control points and measuring points?

_____

48. How does a measuring bridge differ from a universal bench?

_____

_____

Chapter 15 Measurements

Name _____

49. Identify the parts of the measuring bridge shown in the following illustration.

A. _____
B. _____
C. _____
D. _____
E. _____
F. _____

_____ 50. *True or False?* Underbody and upper body measurements are possible with the laser measuring system.

_____ 51. *True or False?* The comparative laser system only works if both sides of the vehicle are undamaged.

52. Name the four features in all computer-aided measuring systems.

_____

_____ 53. Which measuring system uses a bull's-eye?
   A. Ultrasound.
   B. Laser.
   C. Measuring arm.
   D. Gauge.

54. Describe *target bases*.

_____
_____
_____

_____ 55. A(n) _____ is a sketch that shows where damage or misalignment is found on a vehicle.

For questions 56–60, match the following terms with their descriptions.

_____ 56. Distance above datum

_____ 57. Two-dimensional measuring device

_____ 58. Distance from centerline

_____ 59. One-dimensional measuring device

_____ 60. Two diagonal measurements of a hood opening

A. Tram
B. Centerline gauge
C. Width
D. Height
E. X-check

Name _____ Date _____ Class _____

Chapter 16

Unibody Straightening

Learning Objectives

After studying this chapter, you will be able to:
- Identify unibody damage and compare measurements to standards.
- Explain why the unibody must be straightened before structural parts are removed.
- Describe the importance of multiple pulls.
- Identify where to make tie-downs on a unibody.
- Demonstrate how to make hook ups on a unibody.
- Describe the different types of pulls.
- Remove various types of unibody damage.
- Relieve stress in a damaged unibody.

Carefully read Chapter 16 of the textbook and then answer the following questions in the space provided.

_____ 1. The damaged structural parts of the unibody, even if they will be replaced, must be straightened to within _____ of factory specifications.

_____ 2. *True or False?* Kinked high-strength steel parts do not need to be replaced because kinks do not weaken the grain structure of the steel.

_____ 3. *True or False?* Unibody straightening involves the gradual removal of many small stresses in the damaged unirails or structural parts.

4. Why can damage be found far from the point of impact in a unibody?

_____
_____
_____

5. List what must be done to return a distorted unibody back into proper alignment.

_____
_____
_____

6. Name the steps for repairing a damaged unibody.
   _____
   _____
   _____

7. What does a tie-down do?
   _____
   _____

_____ 8. A(n) _____ attaches the pull to the vehicle.

9. What do you do if the hood latch was pushed back and will not release?
   _____

10. Where do you start to locate subtle damage?
    _____

11. Give an example of subtle damage.
    _____
    _____

_____ 12. A rearrangement of the regular pattern in the grains of the steel is known as _____.

_____ 13. *True or False?* A bend causes a more drastic rearrangement of the grain pattern in high-strength steel than a kink.

_____ 14. *True or False?* Treat all structural panels as if they are made of high-strength steel.

15. At minimum, what measuring equipment is required to locate damage in a unibody?
    _____
    _____

_____ 16. *True or False?* When repairing a unibody, straighten the end sections first.

_____ 17. *True or False?* Damage should be corrected in the order in which it occurred.

_____ 18. Which part of the end section is corrected first?
    A. Width.
    B. Length.
    C. Height.
    D. None of the above.

Chapter 16 Unibody Straightening

Name _____

_____ 19. When repairing the end section, lengthening the damage corrects ____.
   A. sidesway
   B. sag
   C. mash
   D. None of the above.

_____ 20. When repairing the end section, restoring height corrects ____.
   A. sidesway
   B. sag
   C. mash
   D. None of the above.

_____ 21. When repairing the end section, setting the width to specifications corrects ____.
   A. sidesway
   B. sag
   C. mash
   D. None of the above.

_____ 22. *True or False?* The tie-downs must be stronger than the pull.

_____ 23. Most unibody vehicles are tied down by ____ placed on the bottom of the rocker panels.

24. When pulling damage from a pickup truck, how should the cab be tied down?

_____

_____

_____

_____

_____

_____

_____

_____ 25. Blocking is the equipment used to prevent ____ movement of structural parts during structural repairs.

_____ 26. The pinch weld clamps bolted to the unibody and tied to the frame machine or floor are considered ____.

_____ 27. *True or False?* If a choice is available between two clamps that will do the job, choose the clamp that has narrower jaws.

_____ 28. *True or False?* A narrow jaw clamp may concentrate the pull force in a small area and tear the metal.

_____ 29. *True or False?* Hammering on a severe buckle without lengthening the panel by pulling will cause additional damage.

30. List the two ways to restore grain pattern in damaged metal.
_____

_____ 31. *True or False?* The correct way to remove buckles is to apply pushing tension.

_____ 32. *True or False?* Even if a part is going to be replaced, it should still be pulled into proper alignment.

_____ 33. In unibody repair, the tolerance for many vehicles is plus or minus _____".

_____ 34. *True or False?* Tolerances do not apply to X-checks.

_____ 35. A pull will always attempt to exert a force in a(n) _____ between the origin of the pull and the tie-down.

For questions 36–40, match the following types of pulls with their descriptions.

_____ 36. Used to increase length of a damaged part parallel to the vehicle's centerline.

_____ 37. Two pulls in opposite directions.

_____ 38. Used to increase width at a 90° angle from centerline.

_____ 39. Pull made in between 90° from centerline and parallel to centerline.

_____ 40. Angle and downward pulls, or angle and upward pulls.

A. Angle pull
B. Combination pull
C. Forward pull
D. Side pull
E. Double stretch pull

41. Why is pulling in one direction at a time preferred to pulling in two different directions at one time?
_____
_____

_____ 42. If a pressure gauge is available on the frame repair equipment, try to keep the ram pressure below _____ psi.

_____ 43. *True or False?* Pulling on two panels rather than one is more likely to tear the sheet metal to the unibody.

44. When making a pull, what should you do if the damaged area is not moving when the pulling force is applied and the pressure gauge on the ram indicates 3000 psi?
_____

_____ 45. Continuing to pull without any movement will cause something to break, usually the _____.

_____ 46. _____ means that the damaged panel is moved slightly beyond the correct location to account for settle back.

Chapter 16 Unibody Straightening

Name _____

47. Define *settle back*.

_____

_____

_____

_____ 48. *True or False?* The hydraulic ram pushes in two directions at the same time.

49. What is the purpose of placing the hydraulic ram on blocks of wood?

_____

_____

_____

50. Describe *stress relief*.

_____

_____

_____

_____ 51. *True or False?* Begin spring hammering at the nearest point of the stress.

_____ 52. What type of flame is used to heat a unibody?
    A. Neutral.
    B. Carburizing.
    C. Oxidizing.
    D. None of the above.

53. What is an advantage to using a heat crayon?

_____

_____

_____

_____ 54. When using heat, stay at least _____" away from any part that will be used to splice to.
    A. 1/2
    B. 1
    C. 2
    D. 6

_____ 55. *True or False?* Kinked panels can be heated during repair.

_____ 56. *True or False?* A bent unirail bumper mounting flange requires tie-downs or pulls.

_____ 57. _____ will minimize settle back when pulling force is released.

_____ 58. During straightening, the hookup should be made on the _____ side.

_____ 59. *True or False?* Kinked rails must be replaced.

_____ 60. When unirail mash occurs, the rail must be _____.

61. What happens to the door when an A-pillar is pushed rearward from a frontal collision?
_____
_____
_____

_____ 62. A side impact can push an A-pillar _____.

_____ 63. *True or False?* Pillar damage can be pushed, but not pulled.

64. When hydraulic power is required during A-pillar repair, what should be done to the vehicle?
_____
_____
_____

65. What does *twist* mean?
_____
_____
_____

_____ 66. *True or False?* Twist may result from a front or rear impact that lowers one corner of the unibody.

67. How is center pillar damage similar to A-pillar damage?
_____
_____
_____

_____ 68. *True or False?* To push out the center pillar, the front seats should be removed.

_____ 69. A _____ impact may bend the rocker panel in, shortening its length.
  A. front
  B. rear
  C. side
  D. None of the above.

70. Name two ways to lengthen a shortened rocker panel.
_____

Chapter 16 Unibody Straightening

Name _____

_____ 71. *True or False?* To prevent the rocker panel from breaking away from the floor during the pull, the rocker panel should be shortened as much as possible.

_____ 72. After removing the roof panel, the roof header and _____ will be accessible.

_____ 73. *True or False?* When repairing a twist in the center section of the vehicle, the vehicle should be tied down during pulls, but not pushes.

_____ 74. Pulls are _____ effective when the hook-up is close to the kink.

_____ 75. One method of pulling on the kink is to use a(n) _____ clamp, which is inserted into the frame rail by feeding it through the hole where the bumper mounts.

Name _____ Date _____ Class _____

Chapter 17

Full-Frame Repair

Learning Objectives

After studying this chapter, you will be able to:
- Make the proper tie-downs to repair various types of full-frame damage.
- Make the proper hookups to repair various types of full-frame damage.
- Make the proper pulls to remove various types of full-frame damage.
- Properly relieve stress in a damaged full frame.

Carefully read Chapter 17 of the textbook and then answer the following questions in the space provided.

_____ 1. A lever consists of a bar that rests on a(n) _____, or pivot point.

_____ 2. *True or False?* Frame repairs should not begin until all damage has been identified.

_____ 3. *True or False?* The tolerances for a full-frame vehicle are ±1/2″ (±12 mm) for linear measurements.

_____ 4. Tolerances for full-frame vehicles are _____ for X-checks.

5. What is the benefit of removing parts?
_____
_____

_____ 6. Tie-downs prevent movement of the vehicle so that _____ move the damage and not the entire vehicle.

_____ 7. Tie-downs may be positioned to help correct the damage by creating _____.

_____ 8. A pivot-point tie-down is positioned at the _____.

9. What does a pivot-point tie-down do?
_____
_____

_____ 10. *True or False?* A counterpoint tie-down is positioned toward the damage and prevents vehicle movement.

_____ 11. _____ prevents a change in height or acts as a fulcrum when a change in height is needed.

Copyright by Goodheart-Willcox Co., Inc. May not be reproduced or posted to a publicly accessible website.

12. What is used to prevent the frame from moving forward, rearward, sideways, and upward?
    _____

_____ 13. *True or False?* If the frame will be pulled forward only, there is no need to anchor to prevent sideways or rearward movement.

14. List the three types of full-frame tie-downs.
    _____
    _____

_____ 15. *True or False?* If the center of the vehicle is damaged, a tie-down at a cross member is made at each of the four corners of the center section.

16. If an angle pull or side pull must be made to the damaged frame, what else is needed?
    _____

_____ 17. *True or False?* The anchoring chain of a tie-down should be loose before pulling.

18. What happens if one tie-down is tighter than another?
    _____
    _____

19. What would cause a tie-down to become slack during a pull?
    _____
    _____

_____ 20. Which hookup has holes in a bar that are lined up to existing holes in the frame?
    A. Chain wrap.
    B. Chair basket.
    C. Bar.
    D. None of the above.

_____ 21. Which hookup has the frame chain around the frame rail or cross member?
    A. Chain wrap.
    B. Chair basket.
    C. Bar.
    D. None of the above.

_____ 22. Which hookup is looped around a cross member and tied to itself?
    A. Chain wrap.
    B. Chair basket.
    C. Bar.
    D. None of the above.

_____ 23. *True or False?* The T-hook works best in a round hole.

Chapter 17 Full-Frame Repair

Name _____

_____ 24. A(n) _____ hook can be used to make a quick hookup to the frame.

25. How do you prevent a clamp from slipping off a frame rail?

_____

_____

_____

_____ 26. A(n) _____ strap acts like a chain and can be used as a quick way to grab a frame?

_____ 27. In full-frame repair, a grade _____ bolt can be inserted through a reinforced hole to attach a U-strap to the frame.
A. 2
B. 4
C. 6
D. 8

28. What happens to a C-channel frame when it is wrapped with chain and pulled?

_____

For questions 29–34, match the following pull types with their descriptions.

_____ 29. Level with the floor and parallel to the frame rails.

_____ 30. Level with the floor and not parallel or perpendicular to the frame rails.

_____ 31. Level with the floor and perpendicular to the frame rails.

_____ 32. Perpendicular to the frame and is used to restore height while correcting kickup.

_____ 33. Attempts to restore length, width (centerline), and height in a single pull.

_____ 34. Pull in opposite directions.

A. Side pull
B. Angle pull
C. Forward pull
D. Downward pull
E. Double stretch pull
F. Combination pull

_____ 35. Several pulls working together are called _____ pulls.

_____ 36. *True or False?* When you must make two pulls in different directions, make both pulls at the same time.

37. What is the most critical part of frame repair?

_____

_____

_____

_____ 38. When the pull is made, watch the tie-downs, the vehicle, and the damage for _____.

39. What should you do if something does not move during the pull?

_____

_____ 40. When stress relieving a full-frame, hammer on any of the following *except:*
   A. rivets.
   B. corners.
   C. rail centers.
   D. angles.

_____ 41. Heat causes the steel grains to move about even more than _____, allowing more effective stress relief.

_____ 42. To heat a full-frame, use a(n) _____ tip on the torch for a large flame.

_____ 43. When heating a frame, heat an area _____ times larger than the buckle.
   A. one to two
   B. two to four
   C. four to six
   D. six to eight

_____ 44. *True or False?* Speed cooling of a heated full-frame by applying water or compressed air.

_____ 45. The tendency of the frame to return to its damaged state once tension is released is known as _____.

_____ 46. Begin full-frame repair by straightening the _____ section of the vehicle.

_____ 47. *True or False?* When repairing frame damage, the length of a frame rail must be restored first.

_____ 48. _____ may be needed to compensate for spring back.

_____ 49. *True or False?* Although both box frame rails will have buckles after a side impact, the buckle in the nonimpact rail will be more severe.

50. When repairing buckles, which pulls are made first, the forward or side?

_____

_____ 51. *True or False?* In a side impact repair, make the nonimpact rail side pull first.

_____ 52. To repair a collapsed cross member, make a(n) _____ pull while pushing up on the cross member at the same time.

_____ 53. *True or False?* Do not try to straighten a damaged engine cross member.

Chapter 17 Full-Frame Repair

Name _____

_____ 54. During a collision, the impact force traveling through a vehicle's frame may cause the frame rails to move away from each other, a condition called _____.

_____ 55. _____ is present when a portion of a damaged frame is lower than the specified height.

_____ 56. Kickup is a condition in which a rail is higher than the specified _____.

57. Describe the characteristics of twist.

_____
_____
_____

_____ 58. Diamond damage affects the _____ section of the vehicle.

_____ 59. _____ causes a shortening of rail length.

60. Define *combination damage*.

_____
_____

Name _____ Date _____ Class _____

Chapter 18

Structural Component Replacement

Learning Objectives

After studying this chapter, you will be able to:
- Explain full replacement with new parts on unibody structural panels.
- Explain sectioning procedures with new parts on unibody structural panels.
- Explain sectioning procedures on full-frame structural parts.
- Explain sectioning procedures with salvaged parts on unibody structural panels.
- Explain rustproofing operations.

Carefully read Chapter 18 of the textbook and then answer the following questions in the space provided.

_____ 1. The first step in structural replacement is _____.
                  A. fit up
                  B. disassembly
                  C. planning
                  D. straightening

2. What happens during disassembly?

_____

_____

3. List three ways to open a door that is smashed shut.

_____

_____ 4. *True or False?* All structural components should be aligned before damaged structural parts are removed.

_____ 5. *True or False?* Measuring should not be done before the vehicle is straightened, but rather after straightening is complete.

_____ 6. Even though the damaged parts will be removed for replacement, they must be _____ first.

_____ 7. *True or False?* You should be careful not to overheat damaged panels that will be replaced.

_____ 8. _____ is removing and replacing only the damaged portion of a structural panel.

Copyright by Goodheart-Willcox Co., Inc. May not be reproduced or posted to a publicly accessible website.

123

9. What is the greatest concern when sectioning?
   _____
   _____
   _____

10. What are a unibody vehicle's structural components designed to do?
    _____
    _____
    _____

_____ 11. *True or False?* In a collision, the sectioned panel must crush in the same way as an original panel.

_____ 12. *True or False?* Reinforcements made of advanced high-strength steel can be sectioned.

_____ 13. Which of the following are *not* commonly sectioned?
   A. Fenders.
   B. Center pillars.
   C. Rear frame rails.
   D. A-pillars.

_____ 14. *True or False?* For sectioning to work, the damaged frame must be cut off first.

_____ 15. All of the following should be avoided when sectioning *except*:
   A. external reinforcements.
   B. drive train mounting locations.
   C. flat areas.
   D. crush zones.

16. What should be done if there are no sectioning instructions available, and there is no good sectioning location?
    _____
    _____

_____ 17. *True or False?* All manufacturers allow sectioning of structural parts.

_____ 18. Before a panel is sectioned or replaced, the location of _____ must be determined before the panel can be removed.

19. How do you remove plug welds from a previous repair?
    _____
    _____

_____ 20. *True or False?* When removing welds, grind on the weld only, not the surrounding metal.

_____ 21. Never grind away metal from the _____ panel.

Chapter 18 Structural Component Replacement

Name _____

_____ 22. What is the proper hole size for a plug weld?
   A. 1/9".
   B. 3/16".
   C. 1/8".
   D. 5/16".

_____ 23. Good welding requires clean, _____ metal surfaces.

_____ 24. *True or False?* Weld-through primer should be applied to bare metal in the weld area after joining the panels together.

_____ 25. At minimum, how large should the gap between a weld and foam be?
   A. 1/4".
   B. 1/2".
   C. 1".
   D. 2".

_____ 26. Which of the following is not part of fit up?
   A. Measuring.
   B. Welding.
   C. Alignment.
   D. Positioning.

27. If you do not have a 3-D measuring system, how can you verify that structural parts are properly positioned before welding?

_____

_____

_____

_____ 28. *True or False?* Laminated steel cannot be MIG welded, and its components are joined by adhesives and rivets.

_____ 29. Replacement structural panels are installed with _____ or continuous welds if a MIG welder is used.

30. If MIG plug welding will be used to join the panels rather than STRSW and adhesives and there are 20 factory spot welds, how many plug welds are required?

_____

_____ 31. A(n) _____ is a short, but continuous bead about 3/4" long.

_____ 32. All of the following are examples of structural joints and welds *except*:
   A. lap/spot.
   B. offset butt/stitch.
   C. butt/plug.
   D. lap/stitch.

33. Which joint/weld is used to section a hydroformed aluminum full-frame?
    A. Offset lap/plug and continuous.
    B. Offset butt/stitch.
    C. Lap/spot.
    D. Butt/plug.

34. *True or False?* The butt joint with insert has a butt weld gap equal to two times the thickness of the panel.

35. What type of welding is recommended on an aluminum frame?
    A. Pulse MIG.
    B. Oxyacetylene.
    C. STRSW.
    D. TIG.

36. What size of rivets is used to bond laminated steel?
    A. 3 mm.
    B. 6 mm.
    C. 9 mm.
    D. 12 mm.

37. *True or False?* After welding, any exposed weld-through primer should be removed.

38. How many coats of epoxy primer are applied to treated weld areas?
    A. Two.
    B. Three.
    C. Four.
    D. Five.

39. _____ foam is flexible enough that it will not change the shape of a panel if too much is applied.

40. Pumping too much _____ foam into the cavity may create over-expansion problems and even cause welded panels to separate.

41. Replacement parts can be new or _____.

42. If a cutting torch is used on the metal at the salvage yard, at least _____" of additional material around the cut should be removed before the part is used.

43. *True or False?* If two or more pieces of a front section are welded together, they should be installed as a unit.

44. Name parts of the side section.

_____

_____

_____

45. In clipping, a rear-damaged vehicle is cut at the _____ pillars and across the floor ahead of the front seat.

Chapter 18 Structural Component Replacement

Name _____

_____ 46. The _____ of a repaired unibody vehicle is dependent on all the relatively thin structural panels retaining their designed strength.

_____ 47. *True or False?* The rivets of a replaced radiator support must be drilled out with a drill bit that is smaller in diameter than the center hole in the rivet.

_____ 48. *True or False?* If the radiator support is supplied in individual pieces, only the damaged portions need to be replaced.

49. When replacing a radiator, what happens to spot welds that impede the removal of damaged parts?
_____

_____ 50. *True or False?* An apron adds greatly to the structural strength of the vehicle.

_____ 51. *True or False?* Aprons should always be replaced entirely and never spliced.

_____ 52. When replacing a radiator, apply weld-through _____ to the bare metal surfaces that will be welded.

_____ 53. The position of the shock tower will determine _____ geometry.

_____ 54. *True or False?* Shock towers should always be replaced entirely and never sectioned.

55. List the parts that make up a left or right 3/4 front.
_____
_____
_____

_____ 56. Begin upper frame rail replacement by removing the _____, fender, and other outer parts as needed.

_____ 57. *True or False?* If a used upper rail in installed, it is usually not part of an assembly.

_____ 58. Lower front frame rails can span the distance from the radiator support to a point under the _____.

_____ 59. If the entire lower frame rail will be replaced, remove the _____.

_____ 60. Damaged advanced high strength steel cannot be repaired or _____.

_____ 61. The A-pillar is part of the _____ and is one of the strongest parts of a unibody vehicle.

_____ 62. All components of an A-pillar are spot welded and continuously welded together in a(n) _____.

_____ 63. If the splice in an A-pillar replacement is made between the door hinge mounts or under the lower hinge mount, the _____ is not involved in the replacement.

_____ 64. The outer rocker panel may be part of a door opening or _____.

_____ 65. When the rocker panel is damaged, the _____ is often damaged as well.

66. Identify the rocker panel parts shown in the following illustration.

A. _____

B. _____

C. _____

D. _____

_____ 67. To begin rocker panel replacement, remove the _____.

_____ 68. If both the inner and outer panels of a center pillar need replacement, a(n) _____ joint is recommended.

_____ 69. Begin center pillar replacement by removing the _____ trim from the pillar.

_____ 70. The rear frame rail is sometimes called a(n) _____.

_____ 71. *True or False?* Damaged trunk floors are often difficult to straighten.

72. Name the three types of joints that are involved in clipping.

_____

_____

_____

_____ 73. When clipping, as with other structural panel replacements, the _____ must be measured and straightened before any cutting is done.

_____ 74. *True or False?* Damaged cross members can be left in place during frame straightening.

_____ 75. *True or False?* Any full-frame damage outside the sectioning area must be cut before it is straightened.

Name _____ Date _____ Class _____

Chapter 19

Steering and Suspension

Learning Objectives

After studying this chapter, you will be able to:
- Identify the parts of the steering and suspension systems.
- List the function of the parts of the steering and suspension systems.
- Detect damaged steering and suspension system parts by inspection or measurement.
- Explain how to remove and install various steering and suspension parts.
- Define wheel alignment terms.
- List the steps of a wheel alignment.

Carefully read Chapter 19 of the textbook and then answer the following questions in the space provided.

1. List the two types of steering systems in use today.

   _____

   _____ 2. The steering shaft assembly connects the steering _____ to the
   _____   steering _____.

3. What does a steering gear do?

   _____
   _____
   _____

4. What connects the steering and suspension systems?

   _____

5. Where can the rack-and-pinion assembly be bolted?

   _____
   _____

   _____ 6. In a rack-and-pinion steering system, the steering linkage consists
                              of inner and outer _____.

7. What is the purpose of the tube that connects the left and right bellows on a rack-and-pinion?

   _____
   _____

_____ 8. The _____ converts the rotation of the sector shaft to a back-and-forth motion.

_____ 9. *True or False?* The pitman arm supports one end of the center link, while the idler arm supports the other end.

_____ 10. *True or False?* All parallelogram steering systems have an idler arm.

_____ 11. What is the connection between the center link and the steering arm?
   A. Steering knuckle.
   B. Idler arm.
   C. Pitman arm.
   D. Tie rod.

_____ 12. *True or False?* The power steering pump lowers the pressure of the fluid in the power steering system.

13. Name the three types of springs used in automotive suspension systems.

_____

_____ 14. _____ are longitudinally mounted steel rods.

_____ 15. *True or False?* The piston rod in a MacPherson strut is much larger in diameter than the rod in a conventional shock absorber.

_____ 16. _____ allow suspension components to move without transmitting vibration and noise to the vehicle's frame or body.

17. Define *strut rods*.

_____

_____

_____ 18. In a MacPherson strut front suspension, the strut assembly is bolted to the steering knuckle and the _____.

_____ 19. Suspension systems that have both upper and lower control arms are known as _____ suspension systems.

20. In a solid rear suspension, what connects both rear wheels?

_____

_____ 21. In a solid rear suspension, the control arm is known as the _____ arm.

_____ 22. *True or False?* In a semi-independent rear suspension, each rear wheel can move up and down separately, but there is an axle between the rear wheels.

_____ 23. *True or False?* In an independent rear suspension, each rear wheel moves up and down in unison.

24. What keeps the vehicle level during acceleration and turns?

_____

Chapter 19 Steering and Suspension

Name _____

_____ 25. A(n) _____ is the surface in the hub that the bearing contacts.

26. What are *wheels*?

_____

_____

_____ 27. *True or False?* All vehicles have the same bolt hole pattern.

_____ 28. In a P-metric tire size designation P205 75 R15, the 15 represents the _____.
   A. rim diameter in inches
   B. rim width in inches
   C. tire width in inches
   D. None of the above.

_____ 29. *True or False?* A comparative-pressure system uses the anti-lock brake system (ABS) to determine if the air pressure in one tire is low compared to the other tires.

_____ 30. *True or False?* If a front wheel is moved back, the suspension system is always damaged.

_____ 31. _____ means that under quick acceleration, the vehicle tries to straighten while turning or pull to one side when traveling straight ahead.

_____ 32. *True or False?* A broken engine mount will cause torque steer.

33. Explain *ride height*.

_____

_____

_____

_____ 34. If the ball joint measurements vary by more than _____", there is probably some type of suspension problem.

_____ 35. The ability of the vehicle's rear wheels to track directly behind the front wheels is known as _____.

_____ 36. Which of the following causes a tracking problem?
   A. Misaligned rear axle assembly.
   B. Improper toe on a rear wheel.
   C. Rear wheels that are not perpendicular to the centerline of the vehicle on a vehicle with a solid rear axle.
   D. All of the above.

_____ 37. *True or False?* Any suspension part that is subject to wear should be replaced with a new part.

_____ 38. *True or False?* Do not use heat on any fastener component that will be reused.

39. How many times can a torque-to-yield fastener be reused?

40. Why should you first count the number of threads visible on the tie rod shaft before removing it?

_____ 41. An out of level rack-and-pinion assembly causes _____.
   A. bump steer
   B. torque steer
   C. pull
   D. drag

42. What aids in lining up the pitman arm to the output shaft?

_____ 43. *True or False?* Tap the pitman arm into position on the output shaft with a sledge hammer.

44. Why is it easier to remove the top nut from a MacPherson strut with an impact wrench than a hand ratchet?

_____ 45. The _____ spring contains potential energy and can shoot from the vehicle with deadly force.

_____ 46. A spring _____ must be used to remove a coil spring on a short/long arm suspension.

47. What does a front wheel that is not correctly positioned in the wheel opening indicate?

_____ 48. To remove a steering knuckle in a MacPherson strut front suspension, begin by removing the _____.

_____ 49. Stabilizer bar links connect the stabilizer bar to the _____.

50. What is the first step when removing the steering column?

_____ 51. *True or False?* Before you consider replacing the damaged part, try straightening a damaged hub, spindle, or steering knuckle.

_____ 52. *True or False?* When removing a seal, work around the seal rather than prying from one position only.

_____ 53. Wheel bearing _____ is used to set tapered wheel bearings in place.

Chapter 19 Steering and Suspension

Name _____

_____ 54. *True or False?* Sealed wheel bearings cannot be serviced.

_____ 55. *True or False?* The only tire damage that is repairable is a small hole in the tread area of a tire that causes the tire to go flat.

_____ 56. *True or False?* Always inflate tires to the pressure specifications on the side of the tire.

_____ 57. To prevent vibration, the tire and wheel assembly must be _____.

_____ 58. Side-to-side motion, or wobble, is caused by lateral _____.

_____ 59. A(n) _____ involves checking to see that the rear wheels follow directly behind the front wheels, as well as checking front wheel alignment.

_____ 60. *True or False?* Front-wheel drive vehicles with adjustable rear wheels require a front-wheel alignment.

_____ 61. *True or False?* If there is frame or unibody damage but the wheels are not hit, the vehicle does not need to be aligned.

_____ 62. *True or False?* If the tires are pointing straight, there is zero toe.

_____ 63. The inward or outward tilt of the tire when viewed from the front of the vehicle is known as _____.

64. Define *cross camber*.

_____

_____

_____

_____ 65. The caster angle is the difference between the _____ axis and true _____.

66. Define *cross caster*.

_____

_____

67. What is *setback*?

_____

_____

_____

_____ 68. *True or False?* Caster is measured directly.

For questions 69–73, match the following terms with their descriptions.

_____ 69. A wheel touching the rear of the wheel opening.

_____ 70. Rear wheels directly behind front wheels.

_____ 71. Does not include passengers.

_____ 72. Part of parallelogram steering.

_____ 73. Dampen road shock.

A. Tracking
B. Idler arm
C. Curb weight
D. Setback
E. Shock absorbers

74. Identify the parts of the MacPherson strut assembly shown in the following illustration.

A. _____
B. _____
C. _____
D. _____
E. _____

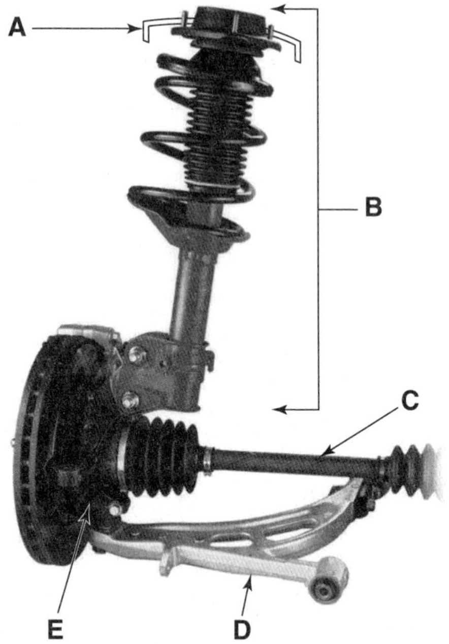

Name _____ Date _____ Class _____

Chapter 20

Electrical System

Learning Objectives

After studying this chapter, you will be able to:
- Identify electrical system components.
- Understand current flow.
- Diagnose an electrical system problem.
- Remove and replace electrical system components.

Carefully read Chapter 20 of the textbook and then answer the following questions in the space provided.

For questions 1–6, match the following terms with their descriptions.

_____ 1. No charge.

_____ 2. Positive charge.

_____ 3. Negative charge.

_____ 4. Measures current.

_____ 5. Measures electrical force.

_____ 6. Measures resistance.

A. Proton
B. Amp
C. Electron
D. Ohm
E. Neutron
F. Volt

_____ 7. A(n) _____ consists of protons, neutrons, and electrons.

_____ 8. When the majority of atoms in a material contain bound electrons, the material is called a(n) _____.

_____ 9. *True or False?* Atoms with free electrons, such as copper or aluminum, are called insulators.

10. How is electricity created?

_____

_____

_____

_____

_____ 11. If a circuit has a continuous pathway, the circuit is said to have _____.

_____ 12. *True or False?* Resistance limits the flow of protons.

_____ 13. Electrons moving through conductors produce _____.

_____ 14. *True or False?* Higher temperatures reduce resistance.

_____ 15. The primary power source in a vehicle is the _____.
　　　A. computer
　　　B. spark plug
　　　C. battery
　　　E. alternator

_____ 16. A(n) _____ circuit has more than one path for the electricity to flow through.

_____ 17. The body of a vehicle serves as the _____.
　　　A. insulator
　　　B. isolator
　　　C. ground
　　　D. inverter

_____ 18. Which of following would make the best ground?
　　　A. Painted metal.
　　　B. Undercoated metal.
　　　C. Plastic.
　　　D. Bare metal.

_____ 19. A broken wire causes a(n) _____.

_____ 20. *True or False?* A bare wire touching the body of the vehicle can create a short circuit.

_____ 21. What causes a bad ground?
　　　A. Loose connections.
　　　B. Corroded connections.
　　　C. Both A and B.
　　　D. Neither A nor B.

_____ 22. *True or False?* Corrosion reduces resistance.

23. List the two components of automotive electrical wire.

_____

_____ 24. The _____ provides the pathway for electrical current.

_____ 25. The _____ prevents electricity from flowing anywhere but through the conductor.

Chapter 20 Electrical System

Name _____

_____ 26. Wires are bundled together in a(n) _____.

_____ 27. A(n) _____ is used to turn on and turn off the flow of electricity.

_____ 28. *True or False?* When a switch is off (open), the circuit is complete.

_____ 29. A(n) _____ is a magnetic switch that uses a low-voltage signal from a remote location to control the on-and-off operation of a high-voltage circuit.

_____ 30. *True or False?* Unrestricted current produces enough heat to melt wires or cause a fire.

31. What is a *fusible link*?

_____
_____
_____

_____ 32. Which of the following is *not* damaged by high current flow?
   A. Fusible link.
   B. Fuse.
   C. Circuit breaker.
   D. All of the above.

_____ 33. *True or False?* Reversible motors operate in two directions by reversing power and ground.

_____ 34. The _____, along with the charging system, is the vehicle's source of electrical power.

35. What does the capacity of the battery measure?

_____
_____

_____ 36. The _____ produces electricity to recharge the battery and provide electricity to various vehicle systems when the engine is running.

37. List the two types of headlights.

_____
_____

38. What are the three functions of taillights?

_____
_____

_____ 39. A(n) _____ governs the operation of the lights.

_____ 40. A car horn is a strong _____ that oscillates a flexible diaphragm made of spring steel.

_____ 41. A(n) _____ is a coil that retracts and expands inside its housing as the steering wheel turns.

42. Name three power accessories on late-model cars.

_____

_____

43. What are the components of a power window system?

_____

_____

_____ 44. A power door lock system uses _____ to lock or unlock the vehicle's doors.

_____ 45. *True or False?* A power mirror system has two motors.

46. What are the components of a power antenna system?

_____

_____

_____ 47. *True or False?* A rear defroster/demister system consists of a conductor grid on the outside of the rear window glass.

48. Describe windshield wiper deicers.

_____

_____

_____

_____ 49. Proper electrical system diagnosis requires _____ reasoning.

_____ 50. *True or False?* Do not replace a blown fuse until the cause of the excessive current is determined and repaired.

51. Name the two types of test lights used to troubleshoot electrical circuits.

_____

_____

_____ 52. *True or False?* When using a nonpowered test light, make sure the vehicle's battery is not charged.

_____ 53. A nonpowered test light consists of a(n) _____ connected to a 12V bulb and ground wire.

54. Describe how to check a fuse in a circuit with a nonpowered test light.

_____

_____

_____

Chapter 20 Electrical System

Name _____

_____ 55. A(n) _____ test light does not need the vehicle battery for power.

_____ 56. *True or False?* A continuity tester cannot be used to check for damaged wiring.

_____ 57. A jumper wire is a length of wire with a(n) _____ at each end.

_____ 58. A jumper wire that lacks a(n) _____ has the potential to carry too much current.

_____ 59. The difference between the voltage on input side and that on the output side of a switch should be less than _____ volts.

_____ 60. *True or False?* Always disconnect power to the switch or relay before testing.

_____ 61. In most wiring diagrams, _____ are used to represent various electrical components.

_____ 62. The joining together of two or more wires by uniting or interweaving the strands is known as _____.

_____ 63. To solder wires together, first slip _____ tubing over one of the wires.

_____ 64. *True or False?* The memory in a clock and radio will be maintained if the battery is disconnected.

65. Identify the components represented by the following symbols.

A. _____
B. _____
C. _____
D. _____
E. _____

_____ 66. *True or False?* A cracked battery case cannot be repaired.

_____ 67. When charging a battery, set the battery charger to _____ amps.
    A. 10–20
    B. 20–40
    C. 40–60
    D. 80–100

_____ 68. A sealed-beam headlight consists of one or two _____, a reflector, and a lens in an airtight assembly.

69. What does accurate headlight aiming determine?
_____
_____

70. What do the V and H values on an aiming tool represent?
_____
_____

_____ 71. If power is not available to the terminal at the horn, check the _____ first.

_____ 72. Collision damage to the front wiper motor is unlikely due to its location on the _____.

_____ 73. To find the cause of a problem when a power accessory will not work, check for voltage at the _____ input.

Name _____ Date _____ Class _____

Chapter 21

Brakes

Learning Objectives

After studying this chapter, you will be able to:
- Understand the function of the brake system and its component parts.
- Identify brake system wear and damage.
- Remove and replace damaged or worn parts.

Carefully read Chapter 21 of the textbook and then answer the following questions in the space provided.

_____ 1. *True or False?* An anti-lock brake system (ABS) helps a vehicle stop faster on dry road.

_____ 2. Brake system operation is based on the principles of hydraulics and ____.

_____ 3. *True or False?* In disc brakes, brake fluid pressure causes the brake caliper to push the brake pads against the brake drum.

_____ 4. According to Pascal's law, if 30 pounds per square inch (psi) of pressure is applied at point A, then ____ psi is also applied at points B and C.

_____ 5. *True or False?* A liquid can be compressed.

_____ 6. *True or False?* The gas produced by boiled brake fluid can be compressed.

_____ 7. To properly function, the brake system must be ____.

_____ 8. ____ opposes the relative movement of two surfaces in contact.

_____ 9. In a drum brake, the brake ____ are pressed against the inside of the brake drum.

_____ 10. In a disc brake, the brake pads are pushed against the brake ____.

11. What are the two groups that comprise brake system parts?

_____

_____

Copyright by Goodheart-Willcox Co., Inc. May not be reproduced or posted to a publicly accessible website.

141

12. Name the friction components of brake systems.
    _____
    _____

_____ 13. Which of the following DOT brake fluids is silicon based?
    A. 3.
    B. 4.
    C. 5.
    D. All of the above.

14. What does the master cylinder do?
    _____
    _____
    _____

_____ 15. A vehicle equipped with a(n) _____ unit is said to have power brakes.

_____ 16. *True or False?* A brake line is a single-walled steel tube.

17. What is the purpose of a brake hose?
    _____
    _____
    _____
    _____

_____ 18. _____ pull the shoes away from the drum when the brake pedal is released.

_____ 19. *True or False?* Nonmetallic linings do not produce as much braking power as semimetallic linings.

_____ 20. The primary shoe is toward the _____ of the vehicle when looking at the drum brake.

_____ 21. *True or False?* The friction material used on all brake pads is semimetallic.

22. Name two things that can cause brake rotors to warp.
    _____

23. What is the purpose of a proportioning valve?
    _____
    _____
    _____

_____ 24. When a truck is heavily loaded, _____ fluid pressure is applied at the rear brakes than when it is not loaded.

_____ 25. The _____ valve senses low brake fluid pressure.

Chapter 21 Brakes

Name _____

26. What limits the hydraulic pressure to the front brakes until a predetermined front input pressure is reached?

_____

_____ 27. _____ occurs when the brakes are applied so hard that the wheels stop turning before the vehicle stops moving.

_____ 28. Which of the following is *not* part of an ABS?
   A. Parking brake.
   B. Tone ring.
   C. Speed sensor.
   D. EBCM.

29. What does the ABS hydraulic actuator do?

_____

_____

_____

_____ 30. Which of the following is *not* a type of ABS?
   A. One-channel.
   B. Two-channel.
   C. Three-channel.
   D. Four-channel.

_____ 31. In a(n) _____ ABS, the rear wheels are controlled together.

_____ 32. *True or False?* The speed sensor for a rear ABS may be located in the transmission, differential, or transfer case.

33. What does rear ABS prevent?

_____

_____ 34. The traction control system (TCS) aids in vehicle _____, not braking.

_____ 35. *True or False?* Specially designed electric motors used in hybrid and electric vehicles reverse direction during braking and become generators that recharge the battery.

36. Describe a flowchart.

_____

_____

_____ 37. Less than ____" of friction material left on the brake pads means they should be replaced.
   A. 1/32
   B. 1/16
   C. 1/10
   D. 1/8

_____ 38. *True or False?* Contamination is the reason many technicians do *not* use penetrating oils when working on brake line fittings.

_____ 39. Some manufacturers require that the _____ be replaced each time the brake line is removed from the caliper.

_____ 40. *True or False?* A rusted or damaged brake line must be replaced.

_____ 41. Be sure to place the flare nuts on the line _____ making the flare.

42. When should brake bleeding be done?
_____
_____

_____ 43. *True or False?* Brake bleeding removes water from the system.

_____ 44. *True or False?* On drum brakes, the bleeder valve is located on the outside of the backing plate.

45. Where is the bleeder valve located on disc brakes?
_____

_____ 46. The pressure bleeder is hooked up to the _____.

_____ 47. In some ABS/TCS systems, a scan tool is required to bleed the _____.

_____ 48. *True or False?* When servicing drum brakes, disassemble one brake at a time.

49. If the brake drum is removed, what happens to the wheel cylinder if the brake pedal is pressed?
_____

50. When replacing a steering knuckle, what needs to be removed?
_____

_____ 51. Which of the following will *not* cause an ABS lamp to indicate a problem?
   A. Debris on the tone ring.
   B. Worn tire treads.
   C. Loose wire on the speed sensor.
   D. Different size tires.

Chapter 21 Brakes

Name _____

52. Identify the different brake system components shown in the following illustration.

A. _____
B. _____
C. _____
D. _____
E. _____
F. _____
G. _____
H. _____
I. _____
J. _____
K. _____
L. _____
M. _____

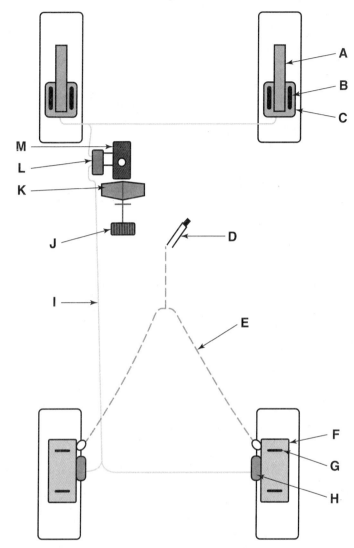

Chapter 22

Cooling, Heating, and Air Conditioning Systems

Learning Objectives

After studying this chapter, you will be able to:
- List the parts of the cooling, heating, and air conditioning systems.
- Explain the function of cooling, heating, and air conditioning system parts.
- Explain how the cooling, heating, and air conditioning systems work.
- Identify damage or wear on cooling, heating, and air conditioning system parts.
- Replace various cooling, heating, and air conditioning system parts.

Carefully read Chapter 22 of the textbook and then answer the following questions in the space provided.

_____ 1. An engine produces heat as it runs. The _____ system manages the heat produced by the engine.

2. When the engine is running the cooling system is pressurized. What does pressurized mean?

_____

_____

_____ 3. _____ is circulated through the cooling system to absorb heat from the engine block.

_____ 4. *True or False?* The coolant is constantly gaining or losing heat.

_____ 5. If the engine is operated with a faulty cooling system, the engine will _____.

_____ 6. The pressure inside the cooling system is about _____ psi.
  A. 14–16
  B. 20–25
  C. 30–40
  D. 100–150

_____ 7. The heat of combustion causes the coolant to expand, creating _____ in the system.

8. Where is the radiator cap on a vehicle with a pressurized coolant tank?

_____

_____

_____ 9. The _____ is the heart of the cooling system.

_____ 10. *True or False?* The water pump is bolted to the engine block.

11. Name two types of hose clamps.

_____

_____ 12. When a cold engine is started, the thermostat is _____.

_____ 13. The _____ acts as a temperature control valve.

_____ 14. Which of the following indicates serpentine belt wear?
   A. 1/2″ deep crack.
   B. missing pieces.
   C. glaze.
   D. All of the above.

_____ 15. What type of collision damage can happen to an electric fan?
   A. Broken mounting tabs.
   B. Bent fan blades.
   C. Crushed motor.
   D. All of the above.

_____ 16. *True or False?* You should not attempt to straighten bent fan blades.

17. Explain how to check for a bent pulley.

_____

_____

_____ 18. *True or False?* Damaged pulleys are not repaired; they are replaced.

19. What do you look for when inspecting a radiator hose?

_____

_____

_____ 20. When checking cooling system hoses for leaks, a _____ spot on the highest point of the hose is the first place to look.

_____ 21. The first step in hose removal is to loosen the _____.

_____ 22. *True or False?* Twisting is not required to remove the hose.

_____ 23. *True or False?* Never position the clamp on the flare of a hose fitting.

Chapter 22 Cooling, Heating, and Air Conditioning Systems

Name _____

_____ 24. What tool measures the specific gravity (relative density) of a liquid?
   A. Hygrometer.
   B. Hydrometer.
   C. Hypometer.
   D. None of the above.

25. What type of water should be mixed with antifreeze?

_____

_____ 26. *True or False?* Antifreeze can be fatal to both humans and animals if ingested. Therefore, coolant should be poured into a container and then disposed of properly.

_____ 27. *True or False?* Flattened radiator fins transfer heat efficiently.

_____ 28. Hot coolant is under _____.

_____ 29. Which type of radiator requires air bleeding?
   A. Radiator above engine.
   B. Radiator below engine.
   C. Both A and B.
   D. Neither A nor B.

30. What happens to trapped air when the cooling system is bled?

_____

_____

_____ 31. Begin electric fan removal by disconnecting the _____.

_____ 32. *True or False?* When the refrigerant absorbs heat, it changes from a liquid to a gas.

_____ 33. When the refrigerant releases heat, it changes from a gas to a(n) _____.

34. Identify the A/C refrigerant phase changes shown in the illustration below.
   A. _____
   B. _____
   C. _____
   D. _____

35. What happens to heat inside the passenger compartment when the air conditioning system is operating?

_____

_____

_____ 36. *True or False?* A change from a liquid to gas, or gas to liquid, is known as a phase change.

_____ 37. *True or False?* Under no circumstances should refrigerant be intentionally discharged into the atmosphere.

_____ 38. *True or False?* The oils in R-12 and R-134a are different and should never be mixed.

_____ 39. The _____ changes the low-pressure refrigerant gas in the system to a high-pressure gas.

_____ 40. When high-pressure gas (refrigerant) leaves the compressor, where does it go to be cooled?

_____ 41. What enters the condenser?
   A. High-pressure gas.
   B. Low-pressure gas.
   C. High-pressure liquid.
   D. Low-pressure liquid.

_____ 42. *True or False?* The condenser is located at the front of the vehicle, behind the radiator.

For questions 43–45, match the following terms to their definitions.

_____ 43. Functions like a radiator to cool the refrigerant.

_____ 44. Removes moisture from the refrigerant and also stores refrigerant.

_____ 45. Changes low-pressure refrigerant gas to a high-pressure gas.

A. Condenser
B. Accumulator or receiver/dryer
C. Compressor

_____ 46. In an air conditioning system, the accumulator must be replaced if the air conditioning system has been opened for more than an hour. Why?
   A. The accumulator lubricant leaks.
   B. When exposed to air, the desiccant in the accumulator will be used up and no longer will be capable of protecting the air conditioner.
   C. Leaking Schrader valve.
   D. None of the above. The accumulator will not need to be replaced.

_____ 47. The process of removing refrigerant from an air conditioning system and storing the refrigerant for reuse is known as _____.
   A. recovery
   B. reclamation
   C. R12 swing
   D. retrieval

Chapter 22 Cooling, Heating, and Air Conditioning Systems

Name _____

_____ 48. *True or False?* The two most common types of refrigerant, R-12 and R-134a, cannot be stored in the same container.

_____ 49. A refrigerant _____ is a machine that is connected to a service fitting. It determines whether the refrigerant is R-12, R-134a, or an unknown refrigerant.

_____ 50. Moisture can cause damaging acids to form in an air conditioning system. Whenever the air conditioning system is opened for service, the air conditioning lines should be _____.
   A. replaced with fresh lines
   B. capped with airtight, plastic plugs
   C. crimped shut with ViceGrip® pliers
   D. cleaned with a long, flexible brush

_____ 51. The _____ is always in front of the radiator and is therefore the most commonly damaged component of an air conditioning system.
   A. condenser
   B. Schrader valve
   C. refrigerant identifier
   D. evaporator

Name _____ Date _____ Class _____

Chapter 23

Power Train

Learning Objectives
After studying this chapter, you will be able to:
- Name and explain the function of power train components.
- Identify damaged power train components.
- Explain how to remove and replace damaged engine and drive train components.

Carefully read Chapter 23 of the textbook and then answer the following questions in the space provided.

_____ 1. A front-wheel-drive vehicle has a(n) _____ mounted engine.

_____ 2. A rear-wheel-drive vehicle has a(n) _____ mounted engine.

_____ 3. *True or False?* In full-time four-wheel drive or all-wheel drive, all wheels are powered at all times.

4. What does an inertia switch do?
_____
_____

5. List the external engine components that may be damaged in a collision.
_____
_____
_____

_____ 6. *True or False?* The intake manifold and plenum are easily damaged in a collision.

_____ 7. To start a vehicle after the inertia switch has been tripped, the switch must be _____.

_____ 8. Which of the following is an indicator of fuel system damage?
A. Leaking fuel.
B. Dented gas tank.
C. Both A and B.
D. Neither A nor B.

Copyright by Goodheart-Willcox Co., Inc. May not be reproduced or posted to a publicly accessible website.

9. What is the purpose of an EVAP system?
_____
_____
_____

_____ 10. Untreated exhaust gases contain which of the following air pollutants?
   A. Hydrocarbons.
   B. Carbon monoxide.
   C. Oxides of nitrogen.
   D. All of the above.

11. How does the air injection system reduce air pollution?
_____
_____
_____

_____ 12. Belts and pulleys are used to transfer motion from the engine's _____ to the power steering pump, air conditioner compressor, and alternator.

_____ 13. *True or False?* Any pulley that wobbles is damaged.

_____ 14. *True or False?* A slightly damaged pulley can often be repaired.

_____ 15. Motor _____ connect the engine to the vehicle's frame or unibody.

16. How can a collision damage a motor mount?
_____
_____
_____
_____

_____ 17. The _____ system channels waste combustion gases from the engine to the rear of the vehicle.

_____ 18. The exhaust system component that burns up polluting exhaust gases is the _____.

_____ 19. *True or False?* The oxygen sensor sends sound waves to the engine electronic control module to help maximize fuel economy and minimize air pollution.

_____ 20. Which of the following are ways that exhaust pipes are joined?
   A. Bolted flanges.
   B. Muffler clamps.
   C. Welding.
   D. All of the above.

_____ 21. To prevent harmful waste gases, such as carbon monoxide, from entering the vehicle, the entire exhaust system must be _____.

Chapter 23 Power Train

Name _____

_____ 22. Hybrid vehicles are being discussed. Technician A says a full hybrid can be powered by the electric motor only. Technician B says a full hybrid can be powered by both the electric motor and the gas engine at the same time. Who is correct?
A. A only.
B. B only.
C. Both A and B.
D. Neither A nor B.

_____ 23. The traction motor uses _____ current to power the front wheel through the transaxle.

24. What is the only way to charge a conventional hybrid?

_____

_____ 25. The high-voltage cables in a hybrid vehicle are _____ for easy identification.
A. green
B. red
C. yellow
D. orange

_____ 26. *True or False?* A hybrid vehicle battery pack does not have a main disconnect switch to shut down the electrical system.

27. Identify the drive train components shown in the following illustration.

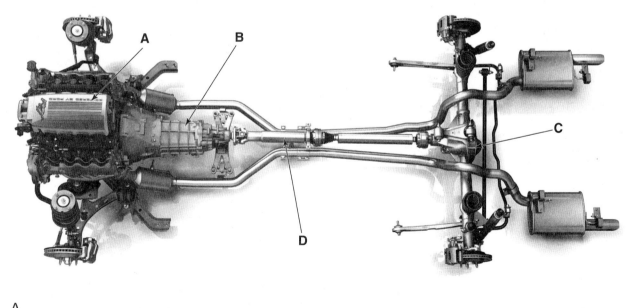

A. _____
B. _____
C. _____
D. _____

_____ 28. A(n) _____ is used in a manual transmission to engage and disengage power from the engine to the transmission.

29. What is the difference between a mechanical clutch and a hydraulic clutch?
_____
_____
_____
_____

30. What does a torque converter do?
_____
_____

_____ 31. *True or False?* The drive shaft must be balanced to prevent vibration.

_____ 32. Which of the following allow the drive shaft to rotate while the suspension moves up and down?
   A. U-joint.
   B. Differential.
   C. Torque converter
   D. Cradle.

33. What is the function of the differential gears?
_____
_____
_____

34. List the advantages of an independent rear suspension over a conventional rear suspension.
_____
_____
_____

35. How do you check for a bent rear axle?
_____
_____
_____
_____

_____ 36. *True or False?* The transaxle does the same job as the transmission and differential in a rear-wheel-drive vehicle.

_____ 37. *True or False?* The drive axles on a four-wheel-drive vehicle are always the same length.

38. What does a CV joint do?
_____
_____

Chapter 23 Power Train

Name _____

_____ 39. *True or False?* The CV joint at the wheel end of the drive axle is called the outer CV joint.

_____ 40. Front-wheel-drive vehicles may be equipped with a(n) _____, which supports the engine, transaxle, front suspension, and rack-and-pinion assembly.

_____ 41. On a four-wheel-drive vehicle, a(n) _____ powers the front wheels through a drive shaft, a front differential, and axles.

_____ 42. In a(n) _____-drive vehicle and a full-time four-wheel-drive vehicle all four wheels are powered at all times.

_____ 43. *True or False?* Some automatic transmissions and transaxles do not have dipsticks.

_____ 44. Which of the following statements about fuel system components is true?
  A. Leaking fuel or dents in the fuel tank are reliable indicators of fuel system damage.
  B. Damaged fuel system parts are generally replaced, not repaired.
  C. Both A and B.
  D. Neither A nor B.

_____ 45. *True or False?* All exhaust hangers can be removed.

_____ 46. What maintains the correct pressure on a serpentine belt?
  A. Belt tensioner.
  B. Belt differential.
  C. Pressed-in-place pulley.
  D. Deep-groove pulley set.

_____ 47. *True or False?* A collision with enough force to bend a pulley may also damage the component behind the pulley.

_____ 48. *True or False?* Removing the drive train to access the front unirails on a front hit hybrid is a similar operation to the removing and installing a front-wheel-drive engine and transaxle.

_____ 49. Technicians should wear _____ gloves when disconnecting the battery pack on a hybrid vehicle.

50. What should be used to remove a heavy battery pack from a hybrid vehicle or an all-electric vehicle?

_____

_____

For questions 51–55, match the following terms to their descriptions.

_____ 51. Removes combustion gases.

_____ 52. Absorbs fuel vapors and routs them back to the engine.

_____ 53. Burns polluting exhaust gases.

_____ 54. Connects engine to frame or unibody.

_____ 55. Includes air filter.

A. EVAP system
B. Air intake system
C. Exhaust system
D. Catalytic converter
E. Motor mount

Name _____ Date _____ Class _____

Chapter 24

Restraint Systems

Learning Objectives
After studying this chapter, you will be able to:
- List the different types of restraint systems and their components.
- Explain the differences and function of active and passive restraint systems.
- Describe the two types of air bag systems.
- Explain the function of the different impact sensors.
- Diagnose restraint system problems.
- Identify and know how to replace restraint system components.
- Understand the safety procedures and inherent dangers of servicing restraint systems.

Carefully read Chapter 24 of the textbook and then answer the following questions in the space provided.

_____ 1. *True or False?* The object of a restraint system is to keep the vehicle's occupants securely seated during a collision and to prevent occupants from striking interior components such as a steering wheel or window.

_____ 2. _____ prevent the occupants from being thrown from their seats during a crash.

_____ 3. The inflated air bags absorb the movement of the vehicle's occupant during a collision, protecting them from the forces of extreme _____.

_____ 4. The headrest acts as a(n) _____ during a rear collision.

_____ 5. *True or False?* The level of protection afforded by the seat depends on the integrity of the seat mount and the metal framework of the seat.

_____ 6. Which of the following is designed to collapse in a collision?
A. Steering column.
B. Air bag.
C. Windshield.
D. Seat belts.

_____ 7. If the windshield is not properly installed, it can be pushed out during air bag deployment. What effect does a lack of windshield have on the passenger safety?
   A. No effect, the seatbelts provide all the protection.
   B. The air bag cannot properly protect the vehicle's occupants without the windshield acting as a backstop.
   C. It depends on how tall the passenger is.
   D. The windshield is designed to pop out during an accident, so it has virtually no effect on passenger safety.

_____ 8. Seat belts that must be buckled by the seat occupant are called _____ seat belts.

_____ 9. The _____ belt restrains the upper body.

_____ 10. *True or False?* A combination lap belt and shoulder harness is called a two-point seat belt.

11. Identify the parts of the active seat belt system shown in the following illustration.
   A. _____
   B. _____
   C. _____
   D. _____
   E. _____
   F. _____

_____ 12. *True or False?* The supplemental restraint systems are intended to add to the protection the seat belts provide.

_____ 13. What tightens the seat belt during a crash?
   A. Post tensioner.
   B. Pretensioner.
   C. Hypotensioner.
   D. Hypertensioner.

_____ 14. *True or False?* Once a pretensioner with an explosive charge has fired, it cannot be used again.

_____ 15. A(n) _____ seat belt system requires the front seat occupants to fasten only the lap belt.

_____ 16. Which of the following is another name for an air bag?
   A. Inflatable restraint.
   B. Supplemental inflatable restraint.
   C. Supplemental restraint system.
   D. All of the above.

Chapter 24 Restraint Systems   161

Name _____

_____ 17. *True or False?* The occupants must be buckled in place for the air bags to adequately protect them.

_____ 18. What is the minimum speed for air bag deployment?
   A. 5 mph.
   B. 10 mph.
   C. 12.5 mph.
   D. 15 mph.

_____ 19. *True or False?* If a frontal impact is not within 30 degrees of a vehicle's centerline, the front-impact air bags will not deploy.

_____ 20. Rapid deceleration causes the coil spring impact sensor's _____ to be thrown forward.

_____ 21. The front-impact air bags will not deploy unless at least one impact sensor and the _____ sensor detect rapid deceleration and signal the module.

_____ 22. *True or False?* Once the air bag system self-check is completed and the components are determined to be functional, the air bag lamp turns on.

_____ 23. The air bag _____ contains the folded air bag and the inflater substance.

_____ 24. In a driver-side air bag, the inflater substance may be made up of _____ pellets.

_____ 25. *True or False?* During a collision, the occupant should contact the air bag just as the air bag reaches full inflation.

_____ 26. A(n) _____ is used to maintain the electrical connection between the driver-side air bag module and the wiring harness.

_____ 27. A(n) _____ does the job of the impact and safing sensors by detecting the deceleration that occurs during a collision.

_____ 28. *True or False?* Advanced air bag systems are designed to determine the severity of a crash and respond appropriately.

_____ 29. *True or False?* The sensors for the side impact air bags may be in the door, center pillar, roof rail, or quarter panel.

_____ 30. A(n) _____ monitors the roll angle and roll rate of the vehicle.

_____ 31. *True or False?* You should not install used air bag system parts.

_____ 32. Air bag modules should not be heated to temperatures above _____ because air bags are designed to deploy at high temperatures.

_____ 33. To prevent accidental deployment, the air bag system must be _____.

_____ 34. The back-up power supply may last up to _____ minutes.
   A. 10
   B. 20
   C. 30
   D. 40

35. What tool do you use to determine if both charges in a two-stage air bag have deployed?
_____

_____ 36. *True or False?* You should carry an undeployed (live) air bag module pointing away from your body.

_____ 37. To replace a clock spring, the steering wheel should be positioned _____.
   A. to the left of center
   B. to the right of center
   C. straight ahead
   D. Steering wheel position does not impact clock spring replacement.

For questions 38–42, match the following restraint system components to their functions.

_____ 38. Tightens seat belt during a collision.

_____ 39. Turns with steering wheel.

_____ 40. Allows the seat belt to be pulled out to fit the occupant.

_____ 41. Monitors roll angle and roll rate.

_____ 42. Detects deceleration.

A. Clock spring
B. Gyrometer
C. Accelerometer
D. Retractor
E. Pretensioner

Name _____ Date _____ Class _____

Chapter 25

Refinishing Tools and Equipment

Learning Objectives
After studying this chapter, the students will be able to:
- Identify hand and power tools used during refinishing operations.
- Identify the types of spray guns.
- Describe the various types of spray booths: cross-draft, semi-downdraft and downdraft.
- Identify equipment used during refinishing operations.
- Name the parts of a compressed air system.

Carefully read Chapter 25 of the textbook and then answer the following questions in the space provided.

_____ 1. Sanding the surface to level out irregularities is called _____ sanding.

2. Which is harder, a sanding block or a sanding pad?
_____

3. What happens if the sandpaper is not tight on the sanding block?
_____
_____

_____ 4. *True or False?* Although sanding blocks are flexible, sandpaper is not.

_____ 5. The sanding _____ is designed to reach into small, difficult-to-access areas.

_____ 6. Unwanted residue is easily removed with a thin, hard rubber block called a(n) _____.

_____ 7. *True or False?* When removing parts, always protect the paint from accidental damage.

_____ 8. *True or False?* An undercoat applied with a roller will have a rougher surface than a sprayed undercoat.

_____ 9. A rough DA sander can be used as a(n) _____ or as a DA sander.

_____ 10. *True or False?* Adhesive-backed disks are single-use products.

_____ 11. *True or False?* Velcro-backed disks cannot be reused.

_____ 12. *True or False?* All air tools should have two drops of air tool oil added before each day of use.

_____ 13. *True or False?* A soft DA pad makes a wider, flatter featheredge than a firm pad.

_____ 14. A(n) _____ uses air pressure to remove dust and moisture from the vehicle.

_____ 15. A _____ spins a wool or foam pad over a painted surface to rapidly remove surface imperfections and polish the paint to a high gloss.
   A. DA
   B. buffer
   C. spinner
   D. jitterbug

_____ 16. The _____ is a rubber wheel that will rub off a tape stripe or decal without harming the underlying paint.

_____ 17. *True or False?* Spray guns use air pressure to break liquid paint into tiny droplets, creating a sprayable mist.

_____ 18. The process of breaking up a liquid into tiny droplets is called _____.

_____ 19. A conventional spray gun uses what air pressure range?
   A. 10–20 psi.
   B. 20–40 psi.
   C. 30–60 psi.
   D. 40–70 psi.

20. What is the difference between overspray and offspray?

_____
_____

_____ 21. An HVLP spray gun _____.
   A. reduces overspray
   B. minimizes paint usage
   C. reduces air pollution
   D. All of the above.

22. What do the letters *HVLP* stand for?

_____

23. What do the letters *LVLP* stand for?

_____

_____ 24. *True or False?* Paint will not flow through the spray gun if the vent hole is clogged.

25. What moves through the two passageways in a spray gun?

_____

26. List the three air exits in an air cap.

_____

Chapter 25 Refinishing Tools and Equipment

Name _____

_____ 27. The fluid tip and the needle are a(n) _____ set.

_____ 28. *True or False?* A correct ratio of air and paint is required for proper atomization.

_____ 29. *True or False?* The size of the hole in the fluid tip determines the amount of paint that can flow out of the spray gun in a given amount of time.

30. Which requires more atomization air, high-solids paint or non-high-solids paint?
_____

_____ 31. *True or False?* Paint is held in a cup located below the air cap in a gravity-feed spray gun.

_____ 32. What is pressurized in a pressure-feed system?
   A. Cup.
   B. Trigger.
   C. Air cap.
   D. All of the above.

_____ 33. When loaded with paint, which spray gun is lighter?
   A. Gravity-feed.
   B. Siphon-feed.
   C. Pressure-feed.
   D. All are the same.

34. Identify the parts of a spray gun shown in the following illustration.

   A. _____
   B. _____
   C. _____
   D. _____
   E. _____
   F. _____
   G. _____
   H. _____
   I. _____
   J. _____
   K. _____

_____ 35. A(n) ____ is an enclosed area in which vehicles are painted.

_____ 36. *True or False?* Paint mist is explosive.

_____ 37. A spray booth that has an exhaust fan only is a(n) ____ pressure booth.
  A. negative
  B. positive
  C. neutral
  D. None of the above.

_____ 38. What does a manometer measure?
  A. Air movement.
  B. Air temperature.
  C. Air pressure.
  D. None of the above.

39. What does an air makeup unit do?

_____

_____

_____ 40. In a(n) ____ spray booth, air moves in through one end of the booth and flows out through the other end.

_____ 41. In a(n) ____ spray booth, the air enters through the roof and is exhausted at or near the floor.

For questions 42–46, match the following terms with their descriptions.

_____ 42. PSI

_____ 43. CFM

_____ 44. Solvent recycler

_____ 45. Spray gun cleaner

_____ 46. UV dryer

A. Enclosed unit that pumps thinner through spray gun.
B. Like a tanning bed.
C. Volume of air.
D. Causes thinner to evaporate.
E. Pressure of air.

Name _____ Date _____ Class _____

Chapter 26

Refinishing Materials

Learning Objectives

After studying this chapter, the students will be able to:
- Understand the difference between thermoplastic paint and thermoset paint.
- List the types of paint commonly used to refinish vehicles.
- Describe the components of paint.
- Understand crosslinking and pot life.
- Explain the uses of primer, surfacer, sealer, basecoat, clearcoat, single-stage, and tri-coat.
- Identify the types of sandpaper used in refinishing.

Carefully read Chapter 26 of the textbook and then answer the following questions in the space provided.

_____ 1. Paints that dry by evaporation of solvents only are called _____ paints.

_____ 2. _____ is the process in which the paint film combines with oxygen.

3. What is polymerization?

_____

_____

_____ 4. *True or False?* The most commonly used paints in collision repair shops today are thermoset paints.

_____ 5. UV cure paints are single-part paints that are available as undercoats and _____.

_____ 6. *True or False?* Powder coating is applied to an entire vehicle.

_____ 7. Solvent-borne paint _____.
  A. contains ether
  B. contains VOCs
  C. causes air pollution
  D. All of the above.

_____ 8. *True or False?* Waterborne paint contains more volatile solvent than solvent-borne paint.

_____ 9. *True or False?* When applying roll-on paint on the undercoat, the roller should be moved in one direction only.

_____ 10. The _____ is the backbone of the paint.

_____ 11. Dried paint is referred to as a paint _____.

_____ 12. What part of paint determines curing characteristics?
　　　A. Binder.
　　　B. Pigment.
　　　C. Solvent.
　　　D. None of the above.

13. Name some binders used in paint.

_____

_____

_____ 14. _____ determine the color of the paint.

_____ 15. _____ is a measure of how well the paint masks the color it is sprayed over.

16. What causes paint to have a metallic effect?

_____

_____

_____ 17. The _____ dissolves the binder and carries the paint.

_____ 18. Which of the following give(s) the paint special capabilities?
　　　A. Solvent.
　　　B. Binder.
　　　C. Additives.
　　　D. Pigment.

19. What causes blistering?

_____

_____

_____ 20. _____ is the measurement of a liquid's thickness.

_____ 21. Lacquer paint is diluted with _____.

_____ 22. Enamel paint is diluted with _____.

_____ 23. Which of the following is the slowest evaporator?
　　　A. Dilution solvent.
　　　B. Adhesion solvent.
　　　C. Leveling solvent.
　　　D. None of the above.

Chapter 26 Refinishing Materials 169

Name _____

_____ 24. *True or False?* Reducer evaporation plays a major role in paint drying.

_____ 25. Each of the following determines the speed of reducer evaporation *except* _____.
A. temperature
B. air flow
C. paint film thickness
D. paint color

26. Does greater airflow through the spray booth filters cause the reducer to evaporate faster or slower?
_____

_____ 27. *True or False?* A low-temperature reducer evaporates slowly.

_____ 28. Paint mixed with a(n) _____ causes crosslinking.

29. Describe *crosslinking* in paint.
_____
_____

_____ 30. Paints that require a hardener are described as two-part or _____ paints.

_____ 31. _____ is the amount of time it will take for the mixed paint to cure under standard conditions of 70°F (21°C) and 50% humidity.

_____ 32. *True or False?* A temperature greater than 70°F (21°C) will increase cure time.

33. What happens if a vehicle is sprayed with paint containing hardener and the vehicle, while curing, is stored at 50°F?
_____
_____

34. Define *shelf life*.
_____
_____

35. What problems may result if a can of hardener has been opened for more than seven days?
_____

_____ 36. Adding a(n) _____ changes the paint by allowing it to remain pliable when deformed.

_____ 37. _____ changes the chemistry of the paint so the paint is attracted to the contamination.

_____ 38. *True or False?* A retarder is used to shorten paint drying time.

Copyright by Goodheart-Willcox Co., Inc. May not be reproduced or posted to a publicly accessible website.

_____ 39. The breakdown of pigment from the UV radiation in sunlight is called _____.

_____ 40. _____ resistance means that the paint is hard enough that it is not easily scratched by abrasion.

_____ 41. The ability to hide slight color variation between the repaired panel and the rest of the vehicle is known as _____.

42. Name the different types of clearcoats.

_____

_____

43. In a tri-coat, how many layers are underneath the clearcoat?

_____

_____ 44. *True or False?* Powder coating produces VOCs and its use will likely decrease in body shops.

_____ 45. *True or False?* A thermoset powder coating should not be reheated to curing temperatures after initial curing.

_____ 46. A(n) _____ is a urethane-, polyurethane-, or epoxy-based polymer engineered to further protect the paint inside the truck bed.

_____ 47. _____ are used to remove microscopic rust formation and to prepare bare steel to accept paint.

_____ 48. To prevent corrosion and to prepare the surface for maximum primer adhesion, a layer of zinc _____ crystals are deposited on the bare steel.

49. What are the two characteristics of a good primer?

_____

50. What is the first line of defense in the fight against corrosion?

_____

51. What type of primer combines cleaning, phosphating, and priming into one procedure?

_____

_____ 52. *True or False?* A wash primer has a vinyl base.

_____ 53. *True or False?* An induction period is the time after mixing when the components in the paint are still interacting.

_____ 54. _____ is the number of minutes that must elapse between coats of paint.

_____ 55. _____ is designed to fill imperfections and is usually applied over the primer.

_____ 56. The appearance of an undercoat pigment in the topcoat is called _____.

Chapter 26 Refinishing Materials

Name _____

_____ 57. _____ is the thickness of paint film, measured in mils.

58. What occurs when improperly mixed surfacer is too thick to flow into and completely fill the scratch?
_____

For questions 59–63, match the following terms with their descriptions.

_____ 59. A glue to hold refinish paint to existing paint.

_____ 60. Primer-surfacer.

_____ 61. Helps achieve a uniform topcoat color and provides hold out.

_____ 62. Can be sprayed and dries to a thick, chip-resistant coating.

_____ 63. Electrode position primer that is applied to a bare metal part at the factory.

A. Rock guard
B. E-coat
C. Adhesion promoter
D. Sealer
E. Dual-use undercoat

_____ 64. *True or False?* To quickly remove paint, 600-grit sandpaper is a poor choice for the job.

65. Name the three classifications for sandpaper.
_____

66. What is the most aggressive tool used in pre-paint preparation?
_____

_____ 67. *True or False?* Chemical paint stripping is a fast and easy way to remove automotive finishes down to bare metal.

_____ 68. Solubility of contaminants is dependent on _____.

_____ 69. Pre-sanding solvent evaporates (quickly, slowly) _____; pre-paint solvent evaporates (quickly, slowly) _____.

70. Describe the purpose of seam sealer.
_____
_____

71. How is liquid mask applied?
_____

72. What is the purpose of buffing materials?
_____
_____

Name _____ Date _____ Class _____

Chapter 27

Paint Mixing and Reducing

Learning Objectives

After reading this chapter, the students will be able to:
- Know where to find a paint code on a vehicle.
- Understand how to use the paint code to obtain the paint part number.
- Understand the importance of thoroughly stirring tints.
- Explain how to mix a color.
- Know how to reduce paint by volume, percentage, and weight.

Carefully read Chapter 27 of the textbook and then answer the following questions in the space provided.

_____ 1. A color formula is a list of the _____ used to make a specific color.

_____ 2. A(n) _____ consists of letters, numbers, or a combination of letters and numbers that represent the color of a vehicle.

_____ 3. *True or False?* The paint code location is the same on all vehicles.

For questions 4–10, match the following vehicle manufacturers to the paint code location(s) on their vehicles. Some answers will be used more than once, and some locations may have more than one answer.

_____ 4. Inside deck lid.

_____ 5. Driver's door edge or pillar.

_____ 6. Console.

_____ 7. Front core support.

_____ 8. Cowl.

_____ 9. Glove box.

_____ 10. Spare tire cover.

A. General Motors (GM)
B. Chrysler
C. Ford
D. Toyota
E. Nissan
F. Honda

11. If a vehicle has a two-tone paint scheme, how many paint codes will be displayed?
_____

12. Why should a technician know the year of vehicle manufacture when looking up a paint code?
_____
_____
_____

_____ 13. A(n) _____ number is the number given to a color by a specific paint manufacturer.

_____ 14. *True or False?* A paint part number will be different for different paint manufacturers.

15. What information do you need to find the paint part number for a vehicle?
_____
_____

_____ 16. One way to find a paint part number is to use a paint manufacturer's _____ book.

17. What could be the cause of the color on some vehicles with the same paint code, model year, and manufacturer to be slightly different?
_____
_____
_____
_____

_____ 18. _____ are variations of a paint color for a specific paint code.
   A. Trackers
   B. Alternates
   C. Rallies
   D. Abstracts

19. Why may a vehicle be a different color under the hood than on the exterior body panels?
_____
_____
_____
_____

_____ 20. Collision repair shops may obtain mixed paint from a supplier called a(n) _____ if they do not have an in-house mixing system.

_____ 21. A(n) _____ is a paint color mixed by the paint manufacturer that may be a better match than a jobber mixed color.

_____ 22. Which of the following items would *not* be considered paint mixing equipment?
   A. Mixing bank.
   B. Paint shaker.
   C. Scale.
   D. Tachometer.

_____ 23. *True or False?* Paint tints can be mixed by weight.

_____ 24. A(n) _____ scale displays a readout as a number rather than a sweeping needle.

Chapter 27 Paint Mixing and Reducing

Name _____

25. What is a simple way a technician can check a digital scale for accuracy?

   _____
   _____

26. What does zeroing a digital scale with an empty paint can on it do?

   _____
   _____
   _____

_____ 27. A(n) _____ is designed to stir tints using an electric motor that rotates stirring bars.

_____ 28. *True or False?* It is acceptable practice to estimate mixing amounts if a technician is only mixing a small amount of paint.

_____ 29. *True or False?* A small amount of high-strength tint will cause a change in color.

30. Why did paint manufacturers develop low-strength tints?

   _____
   _____
   _____
   _____
   _____

_____ 31. Dry powder pearl adds the subtle, glowing effect of _____ flakes to a basecoat.

_____ 32. _____ tint gives a mirror-like effect to a basecoat.

_____ 33. *True or False?* Metallic flakes allow a base color to show through.

_____ 34. *True or False?* Small metallic particles tend to be brighter than larger metallic particles.

_____ 35. _____ is the resin component in paint that forms a film when the paint dries.

36. How is a *single tint paint system* different from a *comprehensive tint paint system*?

   _____
   _____
   _____
   _____

37. List the two jobs a balancer may do in paint.

   _____

   _____

   _____

   _____

_____ 38. A paint formula that lists the weights of each tint required is called a(n) _____ formula.

_____ 39. A(n) _____ formula lists the weight of the entire mixture as each tint is added to the paint.

Use the paint formula shown for Turquoise Green to answer questions 40–44.

Turquoise Green			
Qty: 1 Gallon			
Code	Color	Cumulative	Part
DMC923	Green Sh	308.0	308.0
DMC901	Strong T	746.0	438.0
DMC905	Lemon Ch	1404.8	658.8
DMC900	Strong W	4018.8	2614.0
DMC999	Mixing C	4385.2	366.4
DT885	Reducer	4914.8	529.6

_____ 40. How many grams of the tint Green Sh are added?

_____ 41. How many grams of the tint Lemon Ch are added?

_____ 42. What should the scale read after the color Strong W is added?

_____ 43. What is the last component added for Turquoise Green?

_____ 44. What is this paint manufacturer's paint code for Mixing C?

_____ 45. *True or False?* As a paint mixing guideline, stir tints for 15 minutes twice a day.

_____ 46. Before a new can of tint is placed on a mixing bank, shake the tint for _____ minutes on a paint shaker.

_____ 47. If a tint sits on a shelf for an extended period and ages, it may form _____ that must be double filtered out.
   A. viruses
   B. odors
   C. grit
   D. bubbles

Chapter 27 Paint Mixing and Reducing

Name _____

_____ 48. When mixing paint, add tints to the _____ of the paint can and avoid spreading it out.
   A. center
   B. sides
   C. lid
   D. rim

_____ 49. A(n) _____ occurs when more than the required amount of tint is added.

50. How can some computer-driven scales correct an overpour of tint?

_____
_____
_____

51. What are the two steps of paint reduction?

_____
_____

52. Why must paint be thoroughly stirred or shaken before it can be reduced?

_____
_____
_____
_____
_____
_____
_____

_____ 53. *True or False?* A technician should thoroughly shake water-based paint before use.

_____ 54. *True or False?* Select a fast-drying reducer for high temperature—above 85°F (29.4°C)—spraying conditions.

_____ 55. The amount of time that paint will stay wet enough to accept/absorb overspray is called _____.

_____ 56. Which of the following terms is *not* a way a paint hardener is rated?
   A. Overall-rated.
   B. Flat-rated.
   C. Panel-rated.
   D. Spot-rated.

_____ 57. An indication of paint's liquid thickness is known as _____.

_____ 58. *True or False?* Thin paint has a greater resistance to flow than thick paint, so its flow rate is slower.

_____ 59. Paint viscosity is measured in _____ while using a paint viscosity cup.
   A. gallons
   B. grams
   C. hours
   D. seconds

_____ 60. *True or False?* A viscosity less than the sprayable range means the paint is too thin.

_____ 61. Paint can be reduced by _____, which is measured in parts or percentages.

_____ 62. If 100 ml of reduced paint is required at a 1:1 ratio, _____ ml of paint and 50 ml of reducer are combined.

63. List the volume of each component for 1800 ml of reduced paint with a mixing ratio of 3 parts surfacer, 2 parts reducer, and 1 part hardener.

_____

_____

_____ 64. *True or False?* When reducing paint, the size of the container used will determine the size of each part of the mixture.

_____ 65. A simple _____ stick can be made by using a ruler to mark graduations, or equal units, on a paint stick.
   A. sanding
   B. repair
   C. roller
   D. mixing

_____ 66. *True or False?* Single-use paint mixing cups can save on cleanup time in a body shop.

_____ 67. When reducing paint by percentage, a 300% reduction means that if there is one pint of paint, _____ pints of reducer are added.

68. Briefly explain combination paint volume reduction.

_____

_____

_____

_____

69. Why can't volume reduction ratios be used when reducing paint by weight?

_____

_____

_____

_____ 70. *True or False?* After reducing paint, a technician should leave it alone to allow the paint to combine thoroughly on its own.

Name _____ Date _____ Class _____

Chapter 28

Spray Technique

Learning Objectives

After studying this chapter, the students will be able to:
- Describe the basic skills needed to spray paint. These skills include hand/eye coordination, flexibility, rhythm, and the ability to detect and correct flaws while painting.
- Understand how to set up a spray gun for spray pattern size, material, and air pressure.
- List the spray gun handling variables including: body position, fan orientation, distance, speed, overlap, and triggering.
- Describe the types of coats used when painting panels.
- Demonstrate the procedures for spraying various types of panels.
- Explain how to walk the side of a vehicle.

Carefully read Chapter 28 of the textbook and then answer the following questions in the space provided.

_____ 1. The process of a spray gun using air pressure to break liquid paint into a mist of tiny droplets is known as _____.

_____ 2. _____ is the tendency of the wet paint to level and smooth out after spraying.

_____ 3. *True or False?* Runs are a problem on vertical surfaces because of the downward force of gravity.

4. List four aspects of technician's spray technique that can cause runs.

_____

_____

_____ 5. On a horizontal surface, paint may look rough if the painter does not apply a(n) _____.

_____ 6. On a horizontal surface, wet paint will continue to level and smooth out until it dries due to _____.
   A. internal flow out
   B. external flow out
   C. internal wet coat
   D. internal bond

179

_____ 7. *True or False?* The drier the paint, the longer the leveling time.

8. List the three spray gun adjustments.

_____

_____

_____ 9. The _____ knob regulates the amount of air that flows through the air horns.

_____ 10. The _____ knob adjusts the amount of paint or material that is introduced into the airstream.

_____ 11. *True or False?* A small fan requires less paint; a large fan requires more paint.

12. What results in the paint surface when air pressure cannot break the paint into small droplets?

_____

13. What is the result of insufficient air pressure in a spray gun?

_____

_____ 14. *True or False?* When setting up a conventional spray gun, set the fan first.

_____ 15. *True or False?* When painting a large area, such as a hood, the fan adjustment should be set to make the tall, oval-shaped fan.

16. Why would a technician turn the fluid control knob all the way in and then count the number of turns out?

_____

_____

_____ 17. *True or False?* The air pressure settings on an HVLP spray gun are the same as a conventional spray gun.

_____ 18. The maximum discharge air pressure for the HVLP spray gun is _____ psi.
   A. 2
   B. 4
   C. 8
   D. 10

_____ 19. To calculate pressure, divide force by _____.

_____ 20. *True or False?* If the air pressure on a spray gun is turned down, the fan increases in size.

Chapter 28 Spray Technique

Name _____

21. Explain how to make a pattern test.

_____
_____
_____
_____
_____
_____
_____
_____
_____

22. Identify the cause(s) of each of the pattern test shapes shown in the following illustrations.

   A. _____
   B. _____
   C. _____
   D. _____

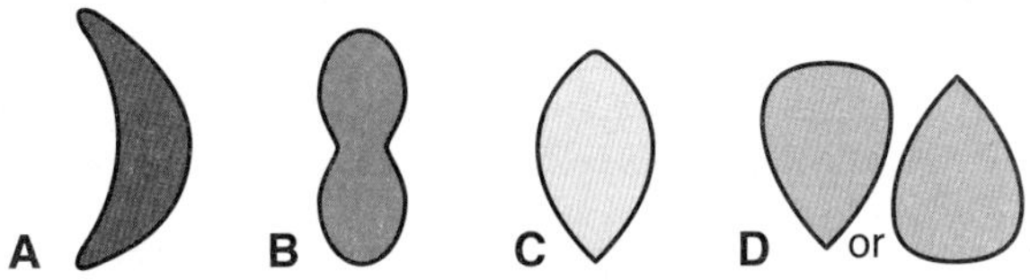

23. What does it mean when a spray pattern is spitting?

_____
_____

24. What is a *flood test*?

_____
_____

25. Explain how to correct each of the flood test shapes shown in the following illustrations.

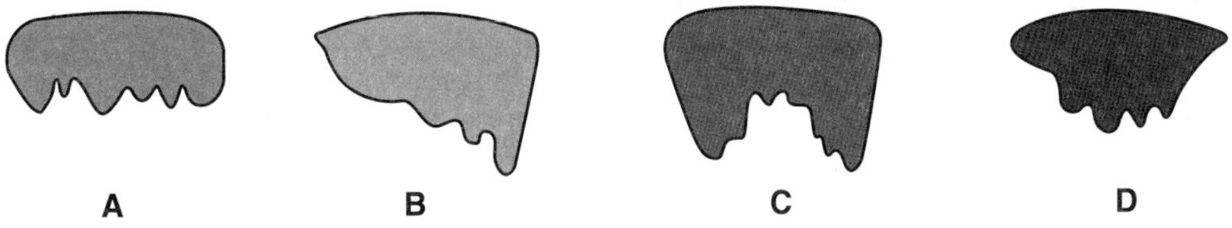

   A. _____
   B. _____
   C. _____
   D. _____

26. List the six spray gun handling areas.

_____
_____
_____

27. What happens if the painter is positioned in the center of the panel rather than the end?

_____ 28. Any fan orientation other than perpendicular to the panel is known as _____.

_____ 29. If the painter does not follow the changing contour of the panel, the resulting problem is called _____.

_____ 30. The travel speed for a conventional gun should be about _____′ per second.

_____ 31. *True or False?* An HVLP spray gun is held farther from the panel and moved slower than a conventional spray gun.

_____ 32. When painting, each stroke should overlap the previous stroke by _____%.
    A. 25
    B. 50
    C. 75
    D. 100

_____ 33. The movement of the gun from one end of the panel to the other is called _____.

_____ 34. Blending calls for a modification of the I-stroke called the _____-stroke.

35. What does *triggering* mean?

36. What is the most commonly used type of paint coat?

_____ 37. _____ occurs when the aluminum flakes in a metallic color are not uniformly distributed.

_____ 38. *True or False?* A drop coat is sprayed at a higher than recommended air pressure setting.

_____ 39. The _____ coat makes a rough surface so the next coats can adhere to it.

_____ 40. A(n) _____ coat is an over reduced basecoat sprayed at one-half normal air pressure.

_____ 41. *True or False?* All paint types are sprayed in the same way.

_____ 42. The method where the painter moves the spray gun down the length of the vehicle is known as _____.

Chapter 28 Spray Technique

Name _____

_____ 43. *True or False?* When spraying a panel, the painter must constantly change body position while spraying.

_____ 44. *True or False?* Painting a horizontal surface such as a hood requires the application of less paint for good internal flow out.

_____ 45. Front end collisions typically require the repair or replacement of the _____ support.

_____ 46. To avoid runs when painting two panels that are at right angles to each other, a spray technique called _____ is used.

_____ 47. *True or False?* Spray with, not against, the direction of airflow in the spray booth.

48. Name two problems that air hoses can present.

_____

_____

_____ 49. *True or False?* Spray guns must be cleaned after every use.

_____ 50. *True or False?* Waterborne paint and solvent paint cannot be mixed together.

_____ 51. *True or False?* Loading a spray gun with thinner and then spraying it out is the best way to clean a spray gun.

Name _____ Date _____ Class _____

Chapter 29

Surface Preparation

Learning Objectives
After studying this chapter, the students will be able to:
- Understand the importance of proper surface preparation.
- Explain how to clean a panel prior to surface preparation.
- Describe various paint removal processes.
- Explain how to mask for primer and paint.
- List the steps in prepping a scratch, body filler, bare metal, weld, rust, repaint, blend, melt, trim, and aluminum.

Carefully read Chapter 29 of the textbook and then answer the following questions in the space provided.

_____ 1. When refinishing a collision repair, the appearance of the topcoat depends on the thoroughness of _____ preparation.

_____ 2. _____ is the ability of two substances to bind together.

_____ 3. When one substance bonds to another through an interaction between atoms or molecules it is called _____.

_____ 4. In _____, the roughness of a surface allows a liquid topcoat to flow into tiny grooves and scratches and "grip" onto the surface.

_____ 5. *True or False?* When preparing surfaces for refinishing, always choose the most aggressive method that will do the job.

_____ 6. The standards for surface preparation can be divided into _____ categories or levels.

_____ 7. *True or False?* Level 1 surface preparation requires the most amount of labor compared to the other levels.

8. During Level 1 preparation, explain what is meant by clearcoat melting.

_____
_____
_____
_____
_____

Copyright by Goodheart-Willcox Co., Inc. May not be reproduced or posted to a publicly accessible website.

185

_____ 9. The goal of Level 3 surface preparation is to make a(n) _____ repair.
  A. noticeable
  B. cheap
  C. undetectable
  D. expensive

_____ 10. Which of the following steps is the first procedure in basic vehicle surface preparation?
  A. Plan the repair.
  B. Inspect the surface carefully.
  C. Paint the area.
  D. Sand the area.

_____ 11. *True or False?* One mil is 0.1".

For questions 12–17, match the following vehicle paint problems with their correct description. Each answer will be used only once.

_____ 12. Appears as reddish brown grains on bare steel.

_____ 13. Gradual change in color due to paint pigment breakdown.

_____ 14. Weather-beaten paint with no gloss due to paint binder breakdown.

_____ 15. Creates a weak, flakey area with bubble-like paint blisters on the surface.

_____ 16. Caused by loss of adhesion between the topcoat and primer.

_____ 17. Begins as tiny slits in the paint surface often caused by UV exposure.

A. Cracking
B. Chalking
C. Peeling
D. Surface rust
E. Fading
F. Internal rust

18. Explain the difference between painted, blended, and melted in reference to how a vehicle panel will be refinished with basecoat and clearcoat.

_____
_____
_____
_____

19. During refinishing, when should a damaged vehicle panel or adjacent panels be blended?

_____
_____
_____
_____
_____

Chapter 29 Surface Preparation   187

Name _____

_____ 20. *True or False?* Repairs of external vehicle panel surfaces may damage internal surfaces.

_____ 21. A separate vehicle panel makes a good edge to end _____ so a repair is unnoticeable.
    A. refinishing
    B. welding
    C. bending
    D. pulling

_____ 22. An entire vehicle should be washed with _____ before starting paint preparations to ensure a quality paint job.
    A. epoxy
    B. soap and warm water
    C. silicone
    D. wax

23. Describe how to remove vehicle molding that is attached with clips.

_____

_____

_____

_____

_____ 24. Which of the following products is used to remove tar and other oil-soluble contaminants?
    A. Pre-sand solvent.
    B. Wax.
    C. Soap and water.
    D. None of the above.

_____ 25. Removing all paint from a surface to leave just the bare metal is known as _____ the surface.

_____ 26. Upon application, a(n) _____ forms a seal on the surface to prevent evaporation of the active ingredients as they break down the paint.

_____ 27. Which of the following materials is *not* an abrasive material that would be used for media blasting?
    A. Sand.
    B. Aluminum oxide.
    C. Plastic beads.
    D. Newspaper shavings.

28. In collision repair, due to the possible secondary damage it can cause, to what use should aggressive media blasting be confined?

_____

_____

_____

_____ 29. When sandpaper stripping paint from a vehicle panel, keep the sander moving to minimize _____ which can melt or damage an area.

_____ 30. *True or False?* When scraping a peeling topcoat with a razor, the angle between the blade and the surface is important for success.

_____ 31. After removing OEM paint on a vehicle with a power sanding tool, a surfacer or sealer will fill in the 320-grit scratches; _____ will not.

_____ 32. *True or False?* A scuffed surface does not require a sealer or surfacer to fill scratches.

_____ 33. Essential for an invisible repair, proper _____ produces a gradual taper of the paint edges around a repair.

_____ 34. As the first step of featheredging, there should be _____" of each paint layer exposed after using 80-grit sandpaper.
   A. 1/16
   B. 1/8
   C. 1/4
   D. 1/2

35. What is the purpose of *back taping*?

_____
_____
_____
_____

_____ 36. Undersurfaces, such as wheel wells, frame, and suspension components, can be protected from overspray with _____.

_____ 37. Which of the following is a rule to follow when masking?
   A. Remove as many obstructions as practical.
   B. Work from inside to outside.
   C. Wash and blow off panels before paint masking.
   D. All of the above.

_____ 38. When bare metal is exposed to air, it will almost instantly begin to _____ as it reacts with the moisture and oxygen in the air.

39. What is the mixing ratio of acid to water for an acid wash used to eliminate flash rust?

_____

_____ 40. *True or False?* Do not spray undercoat over an area that has not been sanded or scuffed, as the undercoat will not stick.

41. What are the two types of primer used in automotive refinishing?

_____
_____

Chapter 29 Surface Preparation

Name _____

42. What problems are caused by an incorrectly mixed thick surfacer?

_____
_____
_____
_____
_____
_____

_____ 43. There are three layers of paint showing in the featheredge. How many coats of surfacer should be applied?
A. 2.
B. 3.
C. 4.
D. 5.

44. How should a technician prep a featheredged surface if the undercoat is exceptionally thick?

_____
_____
_____
_____
_____
_____
_____
_____
_____

_____ 45. *True or False?* Painting improperly cured surfacer will cause shrinkage and sanding scratches.

_____ 46. *True or False?* An infrared dryer cures paint by cooling the surrounding air and lowering the temperature of the entire vehicle.

47. Why does surfacer need to be block sanded?

_____
_____
_____

_____ 48. The absence of guide coat, with no exposed bare metal areas (high spots), indicates a(n) _____ surface.

_____ 49. *True or False?* While sanding surfacer, pressure should be applied to just one end of a sanding stick to properly level the area.

50. Disregarding speed, which type of block sanding, power or hand, better levels surfacer and is more accurate?

_____

_____ 51. During finish sanding, a technician should sand in a(n) _____ motion to remove scratches made while block sanding.
   A. up-and-down
   B. circular
   C. triangular
   D. digging

_____ 52. Keep the vehicle panel surface _____ while finish sanding.
   A. bent
   B. hot
   C. wet
   D. All of the above.

_____ 53. Collecting dust during sanding operations with a(n) _____ sander will lessen surface cleanup time and airborne particles in a shop.

_____ 54. If panels will be painted on the vehicle, which of the following areas is masked for topcoat first?
   A. Panel openings.
   B. Windows.
   C. Wheels and wheel openings.
   D. Adjacent panels.

_____ 55. The _____ of a panel opening will not be directly exposed to overspray as the panel is painted.

_____ 56. Applied to the edge of one panel, _____ fills panel gaps to protect from overspray in the same way as back tape and paper.

_____ 57. Depending on the level of surface preparation, some shops will simply _____ a window molding rather than raise it.

58. Explain how to mask a windshield or back glass opening with masking paper if the glass has been removed.

_____

_____

_____

_____

_____

_____ 59. Which of these items should *not* be used to mask a wheel or wheel opening?
   A. 18″ masking paper and tape.
   B. Single-use plastic wheel coverings.
   C. Towels.
   D. Liquid mask.

Chapter 29 Surface Preparation

Name _____

_____ 60. *True or False?* A body line can be used as a refinish edge if it runs the length of a panel.

_____ 61. Panels adjacent to the spray area should be covered with 18″ ____.

_____ 62. A(n) ____ should be unfolded and gently glided over a surface before paint is applied to remove super-fine contaminants.

_____ 63. ____ is used to fill panel joints and gaps between panels.
A. Body filler
B. Seam sealer
C. Pre-sand solvent
D. None of the above.

_____ 64. Available in an aerosol can, ____ forms a thick, clear, protective layer on the lower surfaces of vehicles.

_____ 65. *True or False?* Self-etching primer can be sprayed on bare aluminum.

_____ 66. *True or False?* If a scratch is too deep to be buffed out, it must be sanded and featheredged.

_____ 67. *True or False?* Pinholes in body filler can be effectively filled with spot putty.

68. What appears in a clearcoat if shrinkage occurs in the area?
_____

69. Describe the steps in surface preparing an appearance weld.
_____
_____
_____
_____
_____
_____
_____
_____
_____

_____ 70. *True or False?* When repairing rust, after grinding, the rust will look like large red rectangles and triangles in the metal.

_____ 71. ____ tapers coats of paint on a vehicle panel, making a gradual transition from refinish paint to existing paint.

_____ 72. If a vehicle panel is _____, the adjacent panels will be blended to better mask the refinish paint.
   A. twisted
   B. crushed
   C. rusted
   D. replaced

_____ 73. A(n) _____ is also known as a solvent blend.

_____ 74. Pre-painting inaccessible areas of new replacement parts prior to installation is known as _____, edging, or cutting in.

_____ 75. *True or False?* An example of a vehicle panel that should be trimmed is the inside of a door.

Name _____ Date _____ Class _____

Chapter 30

Color Matching

Learning Objectives

After studying this chapter, the students will be able to:
- Explain how wavelength absorption determines color.
- Explain why colors appear different under various light sources.
- Describe the qualities of a color: hue, value, and chroma.
- Demonstrate how to view a color.
- Explain how to plot a color.
- Demonstrate how to make a comparison panel.
- Explain what a spectrophotometer does.
- List how spraying factors change colors.
- Explain how to tint a color.

Carefully read Chapter 30 of the textbook and then answer the following questions in the space provided.

_____ 1. _____ is another word for color.

2. List the four primary hues used in automotive refinishing.

_____

_____ 3. *True or False?* All the secondary hues in the color circle are a one-to-one combination of their adjacent primary hues.

_____ 4. _____ is the lightness or darkness of a color.

_____ 5. *True or False?* The value scale has white at the top and black at the bottom.

_____ 6. Light colors have _____ value.

_____ 7. *True or False?* Value is also known as whiteness or blackness.

_____ 8. _____ is the intensity or brilliance of a hue.

_____ 9. *True or False?* Neutral gray has no hue or chroma.

_____ 10. As neutral gray is added to a color, intensity _____.

11. List three other names for chroma.

   _____

   _____

   _____ 12. *True or False?* Adding metallic to a color increases chroma.

   _____ 13. _____ pigment reflects all wavelengths of light.

14. What type of light is used as the standard for color matching?

   _____

   _____ 15. What type of lighting produces light similar to natural light?
   A. Color-corrected.
   B. Incandescent.
   C. Fluorescent.
   D. None of the above.

16. List the two ways colors may be viewed.

   _____

   _____

17. Which has more chroma, solid red or metallic red?

   _____

18. Identify the ways to view a color shown in the following illustrations.

   A. _____
   B. _____
   C. _____

   _____ 19. Which of the following is translucent?
   A. Mica flakes.
   B. Metallic flakes.
   C. Both A and B.
   D. Neither A nor B.

20. List the layers of a tricoat.

   _____

Chapter 30 Color Matching

Name _____

_____ 21. Which of the following is a reason for the refinish color *not* to match the vehicle color?
   A. Vehicles are painted at the factory with a type of paint that is different from the type of paint available to refinishers.
   B. Vehicle manufacturers buy paint from many different manufacturers, and each manufacturer has its own formula for each particular color.
   C. The environmental and spraying conditions are different at each vehicle factory and may vary from one day to the next.
   D. All of the above.

_____ 22. In red-green _____, an individual cannot see a difference in color between a green object and a red object.

23. What are the two most difficult parts of color matching?

_____

_____

_____

_____ 24. *True or False?* A hue can be influenced by adjacent hues.

_____ 25. The subtle presence of an adjacent hue in a color is called _____.

26. What color can be described as being too yellow or too blue?

_____

_____ 27. *True or False?* Always describe the refinish paint in comparison to the vehicle color.

_____ 28. *True or False?* Low value colors may have a slight amount of any hue.

29. What causes paint to fade?

_____

_____ 30. _____ is a condition in which the comparative lightness or darkness of a color changes as the viewing angle changes.

31. What causes flop?

_____

_____

32. List the four characteristics that should be considered when comparing refinish paint to vehicle color.

_____

_____

_____

_____

Copyright by Goodheart-Willcox Co., Inc. May not be reproduced or posted to a publicly accessible website.

33. How much paint is mixed to spray a comparison panel?
_____

34. What is a drawdown bar used for?
_____
_____

_____ 35. A letdown panel is used to check color match of _____.
   A. single-stage
   B. basecoat/clearcoat
   C. tricoat
   D. None of the above.

_____ 36. Color _____ is like making a road map.

_____ 37. _____ is the color of the tint if it were sprayed onto a test panel.

_____ 38. A(n) _____ is an electronic device that determines the color of a vehicle by measuring light intensity (energy) reflected from a sample.

_____ 39. *True or False?* In additive tinting, a specific tint or tints are not added to the existing refinish paint.

_____ 40. _____ is a condition in which two colors look the same under one type of light but look different under another type of light.

For questions 41–45, match the following terms with their descriptions.

_____ 41. Light or dark.   A. Metamerism
                                                                        B. Hue
_____ 42. Color.   C. Value
                                                                        D. Chroma
_____ 43. Color change dependent on viewing   E. Flop
                         angle.

_____ 44. Color appears different depending on
                         type of light.

_____ 45. Intensity.

Name _____ Date _____ Class _____

Chapter 31

Paint Application

Learning Objectives

After studying this chapter, the students will be able to:
- Understand the different levels of surface preparation.
- Recognize sources of contamination.
- Understand how to spray single-stage paint.
- Understand how to spray basecoat/clearcoat paint.
- Understand how to spray tricoat paint.
- Explain the term hiding.
- Understand how to make a blend.
- Understand how to make a melt.
- Recognize and know how to correct common refinish problems.

Carefully read Chapter 31 of the textbook and then answer the following questions in the space provided.

1. List four contaminants found in the shop environment.
   _____

   _____ 2. *True or False?* Color sanding and buffing will remove most contamination, but it is better to prevent contamination from occurring in the first place, rather than repairing it later.

3. What is the greatest source of paint contamination?
   _____

   _____ 4. *True or False?* Blow off the vehicle with compressed air outside the spray booth before masking.

5. When blowing off, what should you pay special attention to?
   _____
   _____

6. A static electrical charge on the vehicle may attract charged contaminants into the wet paint. How do you prevent the buildup of static electricity?
   _____
   _____
   _____

_____ 7. Which of the following is a potential source of contamination?
   A. Vehicle.
   B. Painter.
   C. Spray gun.
   D. All of the above.

_____ 8. When masking a vehicle with masking paper, _____ should always be taped over.

_____ 9. Disposable paper filters may shed _____ into the paint as it is poured through the filter.

_____ 10. _____ paint is color only.

_____ 11. *True or False?* In spot painting, only a portion of the panel is painted.

12. Identify the basecoat blend types shown in the following illustrations.

A. _____
B. _____

13. How does a melt hide overspray?

_____
_____

_____ 14. A(n) _____ coat is a dry, slightly rough coat.

_____ 15. _____ is the number of minutes it takes for the solvents in freshly applied paint to evaporate.

_____ 16. Which of the following may result from *not* allowing adequate flash time?
   A. Runs.
   B. Solvent pop.
   C. Both A and B.
   D. Neither A nor B.

_____ 17. When metallic paint is sprayed, the flakes may not be dispersed uniformly in the paint film. The clumping of the metallic flakes is called _____.

Chapter 31 Paint Application

Name _____

18. Tiger stripes are dark bands in a light metallic color. List three causes of tiger stripes.

_____

_____

_____ 19. A mist coat is sprayed at _____ the normal distance.

_____ 20. *True or False?* Pay attention to the direction of airflow in the spray booth and work with it—not against it.

_____ 21. *True or False?* When painting in a downdraft booth, start spraying at the bottom of the vehicle and work up.

_____ 22. *True or False?* Orange peel results from poor atomization.

_____ 23. *True or False?* Always stop spraying clearcoat in the middle of a panel.

_____ 24. *True or False?* Using a J-stroke will prevent the halo effect.

_____ 25. A back blend is a(n) _____ process.

_____ 26. A(n) _____ panel is sprayed to determine how many midcoats to apply in a tricoat.
A. patch
B. blend
C. let-down
D. match

_____ 27. *True or False?* A tricoat blend requires that both the basecoat and midcoat be stepped out to hide any mismatch.

_____ 28. Which of the following may be used to hide a blend?
A. Contour.
B. Body line.
C. Stripe.
D. All of the above.

_____ 29. *True or False?* A blend requires at least 12″ between the edge of the surfacer or sealer and the edge of the panel, unless there is a panel contour or break to help hide the blend.

_____ 30. _____ is the number of minutes required for the paint to cure to a point where dirt will no longer stick.

_____ 31. *True or False?* Catalyzed paint, such as clearcoat, cannot crosslink and cure at temperatures below 80°F (26.7°C).

32. Which of the following will cause a run?
    A. Moving the spray gun too slowly.
    B. Holding the spray gun too far from the surface.
    C. Not enough overlap.
    D. All of the above.

_____ 33. For sanding a run, _____ is a better lubricant than water.

_____ 34. *True or False?* If a run is made in the clearcoat, you should continue spraying.

_____ 35. _____ are caused by silicone contamination of the paint surface.

_____ 36. The uneven distribution of metallic flake in the paint is called mottling. If mottling is present, what part of the spray gun should you clean to eliminate the problem?
    A. Needle.
    B. Vent hole.
    C. Air cap.
    D. None of the above.

_____ 37. *True or False?* A drop coat is sprayed at twice the normal air pressure, with all other aspects of spray technique remaining the same.

_____ 38. The cure for _____ is to create a(n) barrier between the refinish paint and the incompatible lacquer of thermoplastic paint.

_____ 39. *True or False?* Small solvent pop holes can be removed by sanding after one hour of drying time.

_____ 40. *True or False?* If orange peel is present, adjust the fluid control knob to increase the amount of paint.

_____ 41. *True or False?* Spraying clearcoat is a balance between applying enough material to avoid dry spray while not applying so much that you create runs.

For questions 42–45, match the following terms to their descriptions.

_____ 42. Result of spraying basecoat over a solvent-laden surface.

_____ 43. Caused by silicone contamination of the paint surface.

_____ 44. Caused by improper featheredging.

_____ 45. Caused by high humidity.

A. Blushing
B. Fish eyes
C. Sand scratch swelling
D. Mapping

Name _____ Date _____ Class _____

Chapter 32

Specialty Painting

Learning Objectives

After studying this chapter, the students will be able to:
- Identify an olefin-containing plastic.
- Recognize the importance of obtaining and following the paint manufacturer's recommendations for plastic refinishing.
- Understand why mold release agent must be removed before refinishing.
- List the steps in prepping and painting a new, unprimed, plastic part.
- List the steps in prepping and painting a new, primed plastic part.
- List the steps in prepping and painting a repaired plastic part.
- Know when to use flex agent.
- Explain "custom painting" techniques.

Carefully read Chapter 32 of the textbook and then answer the following questions in the space provided.

_____ 1. *True or False?* Olefin plastics may require surface treatment for proper paint adhesion.

_____ 2. Some _____ plastics may be unpaintable because paint will not adhere to the surface.

_____ 3. *True or False?* Always follow the specific techniques recommended by the paint manufacturer.

_____ 4. All of the following are flexible plastics *except* _____.
   A. TEO
   B. urethane
   C. SMC
   D. TPO

_____ 5. The _____ changes the characteristics of the paint so it will bend without cracking when the flexible part is deformed.

_____ 6. *True or False?* All clearcoats require flex agent.

_____ 7. *True or False?* Parts that require stretching during assembly should be put together before painting.

_____ 8. Plastic primer is also called _____.

_____ 9. A(n) _____ can be performed to determine whether or not a plastic contains olefin.

10. What is the biggest problem in plastic painting?
_____

11. Why is mold release agent added to liquid plastic?
_____
_____

_____ 12. Mold release agent remaining on the surface of the part is a(n) _____ and it will prevent paint from adhering.

_____ 13. *True or False?* Mold release agent is not water soluble.

_____ 14. *True or False?* Wax-and-grease remover should be used to remove oil-soluble contaminants.

15. Identify the tests shown in the following illustrations.

A

B

A. _____
B. _____

_____ 16. Plastic cleaner can be made from a 1:1 mixture of _____ and water.

17. Why should you only clean a plastic surface once?
_____
_____

_____ 18. Prime or paint the part within _____ minutes of sanding.
  A. 5
  B. 10
  C. 30
  D. 60

_____ 19. *True or False?* When spraying plastic parts, spray the difficult-to-reach areas last.

Chapter 32 Specialty Painting

Name _____

_____ 20. *True or False?* Always assume that mold release agent is present on the surface of bare plastic.

_____ 21. Olefin plastics may require the application of a special primer within _____ minutes of sanding.
   A. 5
   B. 10
   C. 30
   D. 60

_____ 22. A primed part will have a(n) _____ surface.

23. Explain how to check for primer adhesion on a primed plastic part.

_____
_____
_____
_____

_____ 24. To check for lifting on primed plastic parts, apply _____ to a rag and touch the rag to the primer.

_____ 25. A(n) _____ is a visible ring around a painted and repaired area.

_____ 26. *True or False?* Applying plastic repair material is allowed by all paint manufacturers.

_____ 27. *True or False?* Basecoat is flexible enough without a flex agent.

_____ 28. *True or False?* Vinyl is painted differently than flexible plastic.

29. What is custom painting?

_____
_____

_____ 30. In a candy color paint job, the base color can be solid or _____.

_____ 31. More coats of pearl tend to reduce the _____ of a color.

_____ 32. What is the overlap percentage for a candy coat?
   A. 25
   B. 50
   C. 75
   D. 100

_____ 33. *True or False?* When spraying candy colors, the painter must strictly follow bodylines.

34. List the three ways to make a design with masking tape.

_____

_____

_____

35. Why can an airbrush atomize paint finer than a normal size spray gun?

_____

_____

36. List two ways to create the illusion of three dimensions on a flat surface.

_____

_____ 37. When fading colors together, the most popular effect is to use _____ colors on the color wheel.

38. What is shaded paint?

_____

_____

_____

_____ 39. Clear _____ is a mixing tint without pigment.

40. How do you make a less defined edge when stenciling?

_____

_____

_____

41. List the two potential problems when stenciling.

_____

_____

_____

For questions 42–45, match the following terms with their descriptions.

_____ 42. The outline of a pattern is formed with masking tape and the paint is applied inside the border of the tape to make the design.

_____ 43. After the paint is allowed to cure, the clearcoat is scuffed and the panel is cleaned.

_____ 44. A color is sprayed and then cured; masking tape is applied over the color to protect the color from overspray.

_____ 45. A design is masked off and the color applied is only slightly different from the background color.

A. Positive masking
B. Negative masking
C. Ghosting
D. Marbleizing

Name _____ Date _____ Class _____

Chapter 33

Detailing

Learning Objectives

After reading this chapter, the students will be able to:
- Identify and remove paint defects, such as runs, dirt, orange peel, single-stage fade, scratches, and acid rain damage.
- Explain how to use a buffer for compounding, polishing, and glazing.
- Describe the procedure for installing stripe tape and moldings.
- Explain how to paint stripes.
- Demonstrate the color sanding process (both dry and wet).
- Describe the procedures for installing decals.
- Detail a vehicle, clean up overspray, and remove oxidation.
- Describe the paintless dent repair process.

Carefully read Chapter 33 of the textbook and then answer the following questions in the space provided.

_____ 1. *True or False?* Detailing is done at the beginning of the painting process.

_____ 2. Sanding a painted surface before buffing is called _____.

_____ 3. *True or False?* Color sanding can remove defects that are under the paint surface, such as lifting, sand scratches, dirt particles in the basecoat, or fish eyes.

_____ 4. Clearcoat must be at least _____ mils thick to protect the underlying basecoat from the harmful effects of the sun's UV rays.
   A. 0.5
   B. 1
   C. 1.5
   D. 2

_____ 5. Clearcoat with dirt or excessive orange peel should first be color sanded with _____-grit sandpaper.
   A. 220
   B. 600
   C. 1000
   D. 1500

6. When hand sanding, what can be used to prevent making finger grooves in the surface?

_____

_____ 7. *True or False?* When power dry sanding, always keep the sander's pad flat.

8. Name three things that should be avoided when color sanding.

_____

_____ 9. The abrasives in a(n) _____ are coarser than those in a polish.

_____ 10. *True or False?* A polish can remove scratches left by 2000-grit or finer sandpaper.

_____ 11. *True or False?* Wool pads run cooler than foam pads.

_____ 12. _____ is a less aggressive buffing process intended to remove swirl marks made by compounding.

13. Explain how a spinning pad works.

_____

_____

14. Identify the conditions shown in the following illustrations.

A

B

A. _____

B. _____

_____ 15. A(n) _____ results when all the clearcoat or paint is removed, exposing the basecoat or undercoat.

_____ 16. When buffing, the buffer must contact all the paint on the surface except the _____.

_____ 17. The buffer travel speed should be about _____′ every 10 seconds.
　　A. 1
　　B. 5
　　C. 10
　　D. 50

Chapter 33 Detailing

Name _____

_____ 18. To buff clearcoat or single-stage topcoat, select a speed of about _____ rpm.
   A. 800
   B. 1000
   C. 1200
   D. 1500

_____ 19. Dried abrasive on the pad can cause _____.

_____ 20. *True or False?* Today's clearcoat urethanes become very soft when fully cured.

_____ 21. A(n) _____ is the short span of time a fresh urethane clearcoat can be easily buffed.

_____ 22. _____ will fill and hide swirl marks.

_____ 23. *True or False?* Wiping a painted surface with dry cloth may cause scratches in the paint if the paint surface or the cloth contains grit or abrasive residue left behind after compounding or polishing.

_____ 24. *True or False?* Paintless dent repair is the best choice if the dent has broken paint.

25. Explain how a light or target board that projects straight lines across a panel helps identify dents.

_____

_____

_____ 26. *True or False?* In paintless dent repair, the pull/push method works best on a bodyline.

_____ 27. *True or False?* When installing pin stripes on the entire side of a vehicle, apply pin stripe to one panel at a time.

28. What may happen if solvents are trapped in the paint by a decal?

_____

_____ 29. *True or False?* If adhesive-attached moldings are reused, all the original adhesive must be removed from the molding.

_____ 30. A quick way to remove adhesive is to heat the blade of a(n) _____ with a propane torch.

_____ 31. New rivets can be welded onto a replacement panel using a(n) _____.

_____ 32. Decals smaller than _____" in diameter are generally installed dry.

33. Why should the technician have clean hands when installing a decal?

_____

_____

Auto Collision Repair and Refinishing Workbook

_____ 34. When detailing the interior, use a(n) _____ to remove remaining dust from the dash, gauges, door panels, and console.

_____ 35. *True or False?* Overspray on glass can be scraped with a razor blade.

_____ 36. A(n) _____ is an engineered resin compound used to remove contaminants from the paint surface and windows.

_____ 37. *True or False?* A lubricating liquid must be applied to the painted surface after using the clay bar.

_____ 38. Dead bugs on chrome, chromed plated plastic, or painted parts can be removed with a(n) _____.

39. What causes orange peel?

_____

_____

_____ 40. To remove dirt particles, use a dual-action sander and _____-grit sandpaper to power sand the dirt spot.

_____ 41. When more paint is applied to a surface than can adhere, the result is known as a(n) _____.

42. Explain how to scrape a run.

_____

_____

_____

_____

_____

43. How can shallow scratches be removed from the clearcoat?

_____

_____

_____ 44. *True or False?* Faded single-stage paint cannot be buffed to bring back the shine.

_____ 45. The pH of rainwater is _____.
   A. 1.5
   B. 3.2
   C. 5.6
   D. 6.8

46. How can acid rain damage be repaired?

_____

_____

_____ 47. To remove acid rain contamination, rinse the vehicle with a(n) _____ wash.

Name _____ Date _____ Class _____

Chapter 34

Estimating

Learning Objectives
After studying this chapter, the students will be able to:
- Understand the importance of customer relations.
- Understand the terminology used in estimate writing.
- Know how to check a vehicle for damage.
- List part sources.
- Know how to find out what is included and not included in labor allowances.
- Know how to calculate judgment times.
- Identify the parts of handwritten and computer-generated estimates.

Carefully read Chapter 34 of the textbook and then answer the following questions in the space provided.

_____ 1. A(n) _____ is a blueprint for the repair of a damaged vehicle.

_____ 2. *True or False?* Estimates are subjective; two shops may evaluate the same vehicle and arrive at different repair totals and may even specify different repair parts.

_____ 3. *True or False?* The estimator is a vital and arguably the most important employee in the shop.

_____ 4. The first step in writing an estimate, handwritten or computer-generated, is to _____.

5. List the information that should be obtained from the customer before the estimate is written.
_____
_____

_____ 6. The estimator has to _____ the repair to the customer.

_____ 7. Damage to a vehicle is called a(n) _____ by an insurance company.

8. When can a liability claim be filed?
_____
_____

Copyright by Goodheart-Willcox Co., Inc. May not be reproduced or posted to a publicly accessible website.

9. When can a comprehensive claim be filed?
   _____
   _____
   _____

_____ 10. *True or False?* The insurance company may ask the vehicle owner to drive the damaged vehicle to at least three collision repair shops for estimates. Or, the owner may be instructed to take the damaged vehicle to a drive-in claims office.

11. List the classifications of vehicle damage.
    _____
    _____

_____ 12. Some suspension fasteners are designed for one-use only and are called _____ bolts.

13. Explain how to center a steering wheel.
    _____
    _____
    _____
    _____

_____ 14. The steering wheel is centered but one front wheel is not pointing straight ahead. Which of the following is most likely damaged?
   A. Steering arm.
   B. Steering gear box.
   C. Pitman arm.
   D. Rack-and-pinion.

_____ 15. The repair cost of a(n) _____ loss vehicle is close to or exceeds the value of the vehicle.

For questions 16–20, match the following terms to their abbreviations.

_____ 16. R&I.   A. Mechanical
_____ 17. R&R.   B. Remove and replace
_____ 18. D&R.   C. Overhaul
_____ 19. O/H.   D. Disconnect and reconnect
_____ 20. m.   E. Remove and install

_____ 21. If a unirail, which is made of high-strength steel, is kinked and straightening would leave it weakened, the unirail must be _____.
   A. pulled while being heated with an acetylene torch
   B. replaced
   C. overhauled
   D. repaired

Chapter 34 Estimating

Name _____

_____ 22. If the cost of repair is more than ____% of the replacement cost, the damaged part is often replaced.

23. List the four sources of parts.

_____

_____

_____ 24. ____ parts are the same as the parts installed on the vehicle when it was made.

_____ 25. *True or False?* Used mechanical parts, such as McPherson strut assemblies, radiators, and alternators, may have wear due to the age and mileage of the donor vehicle.

_____ 26. ____ means that the insurance company will only pay for a part that is as good as, but not better than, the damaged part.

For questions 27–31, match the following terms to their descriptions.

_____ 27. Like kind and quality.

_____ 28. Includes one center pillar.

_____ 29. Includes two center pillars.

_____ 30. Includes radiator support.

_____ 31. Also called rear section.

A. Front clip
B. Rear clip
C. Rear clip top
D. Side clip
E. LKQ

_____ 32. *True or False?* The insurance company may insist on LKQ parts to save money on the claim.

_____ 33. *True or False?* Aftermarket parts are manufactured by an independent manufacturer, not the one that made the OEM parts.

34. List four parts that may be available as remanufactured parts.

_____

_____

35. List two locations where the VIN may be found.

_____

_____

_____ 36. Major parts, which are made up of several smaller parts, are referred to as ____.

_____ 37. *True or False?* There are two types of labor: flat rate for panel repair, and judgment for panel replacement.

_____ 38. ____ is deducted from flat rate labor allowances when replacing one component makes it easier to replace another component.

_____ 39. *True or False?* Price fixing, a conspiracy between body shops to sell their services at a fixed price, is illegal.

_____ 40. A(n) _____ labor allowance is the estimated amount of time given in hours and tenths of an hour, required to perform a specified replacement or refinish task.

_____ 41. Paint labor allowances include the time needed to mask adjacent panels up to _____″ away from the refinish panel.
A. 6
B. 12
C. 18
D. 36

_____ 42. *True or False?* Labor allowances in a flat rate manual are law and must be followed without question.

43. List the formula for calculating the repair judgment labor allowance for steel or aluminum panels.

_____

_____

44. How big is a unit of damage?

_____

_____ 45. *True or False?* Flat rate manuals do not list labor allowances for aftermarket accessories.

46. What is additional labor?

_____

_____

_____ 47. Pricing in a body shop is set up to make a(n) _____% profit on all parts and materials sold to the customer.
A. 15
B. 25
C. 35
D. 50

_____ 48. Body shops are required by law to properly collect, store, and dispose of _____ and keep a record of compliance.

_____ 49. Any additional parts or labor not on the original estimate is considered a(n) _____.

_____ 50. *True or False?* The goal of computer-generated estimates is to minimize effort by having the computer calculate parts and labor totals, figure overlap, and remind the estimator to include certain operations.

Name _____ Date _____ Class _____

Chapter 35

ASE Certification

Learning Objectives
After studying this chapter, the students will be able to:
- Explain the value of ASE certification.
- Understand how to answer ASE style questions.
- List the steps to become ASE certified.

Carefully read Chapter 35 of the textbook and then answer the following questions in the space provided.

_____ 1. _____ certification is a nationally recognized symbol of competence.

_____ 2. A collision repair shop that displays the _____ logo employs technicians that are knowledgeable, experienced professionals.

_____ 3. *True or False?* The National Institute for Automotive Service Excellence, or ASE, is a nonprofit organization that offers a voluntary testing program for automotive technicians.

_____ 4. A technician who passes an ASE test and has the required _____ is awarded ASE certification.

_____ 5. *True or False?* Anyone can call themselves a collision repair technician, whether they possess the skills and knowledge to make acceptable repairs or not.

6. ASE was founded in what year?
_____

7. About how many automotive technicians hold ASE certification?
_____

_____ 8. Which of the following ASE tests is for mechanical and electrical components?
A. B2.
B. B3.
C. B4.
D. B5.

Copyright by Goodheart-Willcox Co., Inc. May not be reproduced or posted to a publicly accessible website.

_____ 9. Which of the following ASE tests is for structural analysis and damage repair?
   A. B2.
   B. B3.
   C. B4.
   D. B5.

_____ 10. Which of the following ASE tests is for non-structural analysis and damage repair?
   A. B2.
   B. B3.
   C. B4.
   D. B5.

_____ 11. Which of the following ASE tests is for painting and refinishing?
   A. B2.
   B. B3.
   C. B4.
   D. B5.

_____ 12. *True or False?* An individual who passes all five collision repair and refinishing tests (B2, B3, B4, B5, and B6) and has two years of relevant, full-time work experience is certified as an ASE master collision repair/refinishing technician.

_____ 13. Certification means that a school meets the NATEF requirements for _____, _____, and _____.
_____
_____

_____ 14. *True or False?* A school that has NATEF accreditation follows guidelines that are proven to develop competent technicians.

_____ 15. *True or False?* ASE does *not* promote education to develop standards for automotive repair training programs.

16. What do the letters *ASE* stand for?
_____

17. What do the letters *NATEF* stand for?
_____

_____ 18. NATEF has established a list of _____ that serve as a solid base for course-of-study outlines and accurately describes the school's training programs.

19. Who develops the NATEF task list?
_____
_____
_____

Chapter 35 ASE Certification

Name _____

_____ 20. *True or False?* This textbook was designed to teach the NATEF tasks.

_____ 21. By seeing that the shop's technicians are ASE _____, the customer will feel more confident that repairs will be made correctly.

22. How much hands-on full-time experience is required for ASE certification?

_____

_____ 23. *True or False?* Part-time experience counts toward ASE experience requirements.

_____ 24. *True or False?* Three years of high school collision repair training may be substituted for one year of the hands-on work experience.

_____ 25. *True or False?* Two years of post-secondary collision repair training may be substituted for one year of the hands-on work experience.

_____ 26. *True or False?* Completion of an apprenticeship or co-op program may be substituted for the entire two years of hands-on work experience.

27. What is the name of the form you use to list your work experience?

_____

28. What are the two things you must do to be ASE certified?

_____

_____

_____ 29. *True or False?* The Work Experience Reporting Form is available online.

30. What happens if you pass an ASE test but do not have the required experience?

_____

_____

_____

_____ 31. Every ASE test question contains one correct answer, _____ incorrect answers, and enough information to choose the right answer.

32. What is the address for the ASE website?

_____

_____ 33. _____ tests and study guides are available from various publishers to help prepare for the exam.

_____ 34. *True or False?* Books, calculators, and other reference materials are permitted in the test room.

_____ 35. ASE certification is good for _____ years.

_____ 36. When ASE certification expires, the technician must take a(n) _____ test to remain certified.

_____ 37. *True or False?* The recertification test has about twice the number of questions of a regular certification test.

_____ 38. Recertification is good for _____ years.

_____ 39. All questions on ASE certification tests are _____ questions.

_____ 40. *True or False?* The Technician A, Technician B type of question can be thought of as a combination of two true or false questions.

_____ 41. A(n) _____ question contains the words *except*, *not*, *least likely*, or *most likely*.

_____ 42. _____ sentence questions are also called fill-in-the-blank questions.

_____ 43. *True or False?* Always read each answer carefully before making your choice.

_____ 44. *True or False?* A technician who does not graduate from an ASE certified education program must have two years of collision repair experience and pass the tests to become certified.

Name _____ Date _____ Class _____

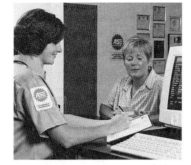

Chapter 36

Employment Strategies and Employability Skills

Learning Objectives

After studying this chapter, the students will be able to:
- Know what a job in a body shop is like.
- Understand the payment methods.
- Describe the job search process, including the interview.
- Understand how to keep a job.
- Know how to obtain training after finding a job.

Carefully read Chapter 36 of the textbook and then answer the following questions in the space provided.

1. What is the most important question of the job search?

   _____

_____ 2. *True or False?* It can be difficult for an individual to know what type of work he/she wants to do while still in school.

_____ 3. True career _____ comes from doing a good job and feeling that the job is important.

_____ 4. In a(n) _____ arrangement, a student spends time observing what goes on in a business, such as a collision repair shop.

_____ 5. *True or False?* Before picking a job or career, explore as few careers as possible to make the best choice.

_____ 6. Although collision repair and refinishing technicians are exposed to harmful chemicals on a daily basis, there are tools, procedures, and technology to dramatically _____ exposure.
   A. magnify
   B. numb
   C. increase
   D. reduce

7. Which payment method uses a specified amount paid to an individual for hours worked only?

   _____

_____ 8. Hours worked beyond eight hours in one day of a 40-hour workweek are called _____ hours.

_____ 9. *True or False?* Overtime hours are often paid at a lower hourly rate.

Copyright by Goodheart-Willcox Co., Inc. May not be reproduced or posted to a publicly accessible website.

_____ 10. Hourly workers usually punch a(n) _____ that records when the employee arrives and leaves.

_____ 11. In a(n) _____ payment arrangement, an individual is paid a certain amount of money each week, regardless of the number of hours worked.

_____ 12. *True or False?* Salary workers are paid for overtime.

_____ 13. _____ means a technician is paid a dollar amount based on each vehicle repaired.

14. If a vehicle estimate states 18 hours of body repair and the shop labor rate is $60 per hour, how much is a collision repair technician, working on 50% commission, paid for the repair?

_____

_____ 15. *True or False?* Commission workers are paid more if the repair takes longer than the estimated time.

_____ 16. *True or False?* An organized, fast-working technician can get paid for 80–100 commission hours of work in a 40-hour workweek.

_____ 17. A(n) _____ is a redo of sloppy work.

_____ 18. *True or False?* The repair work required to fix a comeback is paid at one half the normal commission rate.

_____ 19. *True or False?* A technician working on commission receives no pay if there is no work coming into the shop.

_____ 20. A(n) _____ lumps labor produced by all of a shop's technicians together and pays each technician a percentage of the total.

_____ 21. A collision repair shop using a _____ quota pays a technician a bonus if his/her labor hours match or exceed the quota in a 40-hour workweek.
   A. productivity
   B. parts
   C. vacation
   D. None of the above.

_____ 22. Which of the following is a possible benefit to an employee from an employer?
   A. Paid health insurance.
   B. Vacation time.
   C. Paid retirement.
   D. All of the above.

_____ 23. *True or False?* Many small businesses, which include most collision repair shops, are able to offer benefits for employees.

_____ 24. *True or False?* Usually an employee must work the day before and the day after a holiday to be paid for the holiday.

Chapter 36 Employment Strategies and Employability Skills

Name _____

_____ 25. A(n) _____ means that a technician does not report for work until the employer calls him back.

_____ 26. A technician is not paid during a layoff but may be eligible for _____ based on the circumstances.
A. sick time
B. labor bonuses
C. unemployment benefits
D. vacation time

_____ 27. *True or False?* All job openings are advertised publically.

_____ 28. In the _____ method of job hunting, someone knows about a shop looking for a technician and tells or recommends an interested person.

_____ 29. *True or False?* A résumé left with a shop manager even when the shop is not hiring could provide a job opportunity in the future.

_____ 30. During a job _____, a technician is given the opportunity to learn more about a shop and to convince the employer that he/she is the best person for the job.

_____ 31. What should be the first step in effectively preparing for a job interview?
A. Finalize plans to leave current employment.
B. Research the prospective employer and the job.
C. Pick out an outfit to wear.
D. Decide to accept any employment offer or terms.

32. Before an interview, what should an individual create and answer to mimic the interview process?

_____
_____
_____

_____ 33. Arriving just a few minutes _____ to a job interview creates an awful first impression and can be a deal killer before the process has even begun.

_____ 34. Due to legal aspects, all of the following are topics a prospective employer can discuss with an applicant during an interview *except*:
A. ability to do the job.
B. past work experience.
C. former employers.
D. religious beliefs.

35. List the five work factors to evaluate when comparing two or more job positions.

_____
_____
_____

Copyright by Goodheart-Willcox Co., Inc. May not be reproduced or posted to a publicly accessible website.

_____ 36. One of the best ways to remain employable and in-demand is to continue developing personal work _____.

_____ 37. Instruction, the desire to improve, and _____ will help build skill.
  A. exhaustion
  B. money
  C. experience
  D. cutting corners

_____ 38. A(n) _____ employee is always prompt and on time when the workday starts and when returning from breaks and lunches.

39. What does habitual lateness show to shop owners, managers, and coworkers?
_____
_____
_____

_____ 40. A technician taking initiative and starting tasks without being told is displaying _____, or an inner urge to perform well.

_____ 41. *True or False?* A technician's attitude is not important in determining success on the job.

_____ 42. *True or False?* A 125% production rate would be 50 hours of labor in a 40-hour workweek.

_____ 43. Technicians should recognize that _____ customers keep the collision repair shop open and successful by providing the shop with their business.

_____ 44. If a technician makes a mistake and causes damage, the damage should first be reported to the _____ and not covered up.
  A. secretary
  B. customer
  C. estimator
  D. shop manager

45. What does *ICAR* stand for?
_____

46. What is the job of ICAR?
_____
_____
_____
_____

_____ 47. *True or False?* The information available through Tech Cor is vehicle model specific and thoroughly detailed.

_____ 48. *True or False?* Trade magazines and annual trade shows provide little in terms of new skills, techniques, or information.

Name _____ Date _____ Class _____

Job 1

Safety

Objective
After completing this job, you will be able to recognize and minimize body shop hazards. You will also know where emergency and safety equipment is located.

Equipment and Materials
To complete this job, you will need the following:
- Safety glasses
- Goggles
- Face shield
- Leather gloves
- Ear protection
- Dust mask
- Respirator
- Supplied-air respirator
- Neoprene gloves
- MIG welding helmet
- Welding gloves
- Welding jacket

Safety Notice: Before performing this job, review all pertinent safety information in the *Auto Collision Repair and Refinishing* textbook and discuss safety procedures with your instructor.

Procedure

Proper Safety Practices

1. It is important to understand the hazards encountered in the collision repair shop and to take steps to minimize the dangers associated with them. Explain the difference between chronic and acute hazards and give examples of each.

   _____
   _____
   _____
   _____

   Completed ❑

2. Examine each fire extinguisher in the shop and fill out the following chart.

Location of Fire Extinguisher	Type of Fire Extinguisher	Most Recent Inspection Date

Completed ❑

Job 1 Safety

Name _____

3. Appropriate personal protective gear should we worn when working in the shop. The type of gear worn depends on the task being performed. See **Figure 1-1**. Locate and explain the purpose/use of the following personal safety equipment.

   A. Safety glasses
      Location: _____
      Purpose: _____

   B. Goggles
      Location: _____
      Purpose: _____

   C. Face shield
      Location: _____
      Purpose: _____

   D. Leather gloves
      Location: _____
      Purpose: _____

   E. Ear protection
      Location: _____
      Purpose: _____

   F. Dust mask
      Location: _____
      Purpose: _____

   G. Respirator.
      Location: _____
      Purpose: _____

   H. Supplied-air respirator
      Location: _____
      Purpose: _____

   I. Neoprene gloves
      Location: _____
      Purpose: _____

   J. MIG welding helmet
      Location: _____
      Purpose: _____

K. Welding gloves

   Location: _____

   Purpose: _____

L. Welding jacket

   Location: _____

   Purpose: _____   Completed ❑

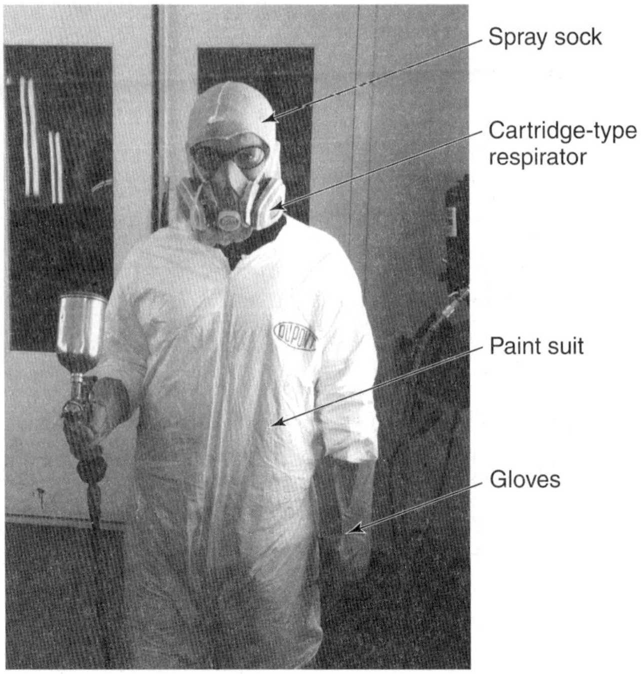

Figure 1-1. The collision repair shop can be a hazardous environment. It is your job to protect yourself at all times. This technician is wearing safety equipment that is appropriate when spraying non-isocyanate paint.

- Spray sock
- Cartridge-type respirator
- Paint suit
- Gloves

4. Where are fire alarms located?

_____

_____   Completed ❑

Job 1 Safety

Name _____

5. Good housekeeping helps minimize safety hazards and increase production. See **Figure 1-2**. Are the work areas in your shop clean and well organized? If not, explain.

_____

_____   Completed ❑

Figure 1-2. Good housekeeping includes the proper disposal of combustible waste. Oily and solvent-soaked rags should be stored in appropriate containers. *(Justrite Manufacturing Company)*

6. Are first-aid kits easy to locate and adequately stocked? If not, list any problems found.

_____   Completed ❑

7. Why is hearing protection important?

_____

_____   Completed ❑

Instructor's Initials _____

Date _____

Name _____ Date _____ Class _____

Job 2

Vehicle Basics

Objective
After completing this job, you will be able to check automotive fluid levels, measure tire pressure, charge a battery, locate a VIN, raise a vehicle on a lift, check for suspension damage, and place a vehicle on jack stands.

Equipment and Materials
To complete this job, you will need the following:
- Unibody vehicle
- Lift
- Jack stands
- Floor jack
- Battery charger
- Tire pressure gauge
- Tape measure

Safety Notice: Before performing this job, review all pertinent safety information in the text and discuss safety procedures with your instructor.

Procedure

Locate and Inspect Parts and Record Information

1. Check the level of the fluids listed in the following chart. Place a checkmark in the chart to indicate if the fluid level is low, okay, or high.

Fluid Type	Fluid Level		
	Low	Okay	High
Engine oil			
Transmission or transaxle fluid			
Engine coolant			
Power steering fluid			
Brake fluid			

Completed ❏

2. Locate and inspect the drive belt. Count the number of cracks on a 1″ length of the belt. Record the number of cracks per inch below. Is the belt in need of replacement?

_____

Completed ❏

3. Locate and inspect the battery. See **Figure 2-1**. Does the battery have top terminals, side terminals, or both?

_____

Completed ❏

Figure 2-1. This battery has top terminals. Virtually every repair job begins with disconnecting the battery. Always check the battery for damage.

4. The battery cable-to-terminal connections should be clean, tight, and in good condition. Did you notice any problems with the cable-to-terminal connections? If so, explain.

_____

Completed ❏

Warning: Always wear safety glasses while working on a battery.

Job 2 Vehicle Basics

Name _____

5. Locate the shop's battery charger. Summarize the procedure for hooking up the battery charger and charging the battery.

   _____
   _____
   _____
   _____

   Completed ❑

6. Find the emission sticker under the vehicle's hood. Write down the engine size, in liters, below.

   _____

   Completed ❑

7. Locate the VIN on the vehicle at hand and write the VIN below.

   _____

   Completed ❑

8. Check the air pressure in each tire. Record the air pressure reading for each tire listed below.

   Right front: _____
   Left front: _____
   Left rear: _____
   Right rear: _____

   Completed ❑

9. Locate the alphanumeric tire size designation on the sidewall of one of the tires and write it below.

   _____

   Completed ❑

10. Explain what each part of the tire size designation found in Step 9 means.

    _____
    _____
    _____
    _____
    _____

    Completed ❑

11. Locate the vehicle lift in your shop. Summarize how to position and raise a unibody vehicle with a lift.

    _____
    _____
    _____
    _____

    Completed ❑

12. With your instructor's help, use the shop lift to raise the vehicle. With the vehicle on the lift, identify the following parts: radiator, condenser, pulleys, engine oil pan, transaxle oil pan, steering rack, tie rod, tie rod end, lower control arm, ball joint, steering knuckle, brake disc, McPherson strut, exhaust system, gas line, front cross member, lower frame rails, rocker panels, and rear frame rails. Completed ❏

13. Find symmetrical reference holes on the lower frame rails. Measure the distance from each lower ball joint to the reference hole on each lower frame rail. Write the measurements below. This is a quick test for suspension damage.

    Right side: _____

    Left side: _____ Completed ❏

14. Wiggle a front wheel to check for play in the ball joints and tie rod.

    Where should your hands be positioned on the tire when checking ball joints?

    _____

    Where should your hands be positioned on the tire when checking the tie rod?

    _____

    Was excessive play in the ball joints or tie rod found? _____ If so, explain.

    _____
    _____
    _____
    _____ Completed ❏

15. Measure the depth of the tread on one of the front tires. Make five measurements and record the average tire thread depth here.

    _____ Completed ❏

16. Examine each tire for wear. Describe the wear patterns found.

    Right front: _____

    Left front: _____

    Left rear: _____

    Right rear: _____ Completed ❏

Job 2 Vehicle Basics 231

Name _____

17. Check the underside of the engine compartment for leaking fluid. Were leaks found? If so, what are the possible sources?

_____
_____
_____
_____ Completed ❏

18. Follow the brake lines from the master cylinder to each wheel. Check the lines for leakage. Did you find any leaking brake fluid?

_____ Completed ❏

19. Where do you find the brake hoses? What was their condition?

_____
_____ Completed ❏

20. Check the gas tank for leakage. Did you find any leaking gas?

_____ Completed ❏

21. Check the pipes, muffler, and other exhaust system components for rust or holes. Did you find any problems?

_____ Completed ❏

22. With your instructor's supervision, lower the vehicle to the ground and drive it off the lift.

_____ Completed ❏

23. Raise the vehicle using a floor jack and secure it on jack stands. See **Figure 2-2**. Summarize how to raise the vehicle with a floor jack and support it with jack stands.

_____
_____
_____
_____ Completed ❏

Figure 2-2. Jack stands are the safe way to support a vehicle raised by a floor jack.

24. List four areas, parts, or structures of the vehicle that should not be used when raising a vehicle with a floor jack or lift.

   1. _____
   2. _____
   3. _____
   4. _____   Completed ❑

Instructor's Initials _____

Date _____

Name _____ Date _____ Class _____

Job 3

MIG Welding Basics

Objective
After completing this job, you will be able to make the basic MIG welds.

Equipment and Materials
To complete this job, you will need the following:
- Hood that is stripped to bare metal or a 3′ × 3′ section of 18–22 gauge bare sheet metal
- MIG welder
- 3″ × 3″ square of sheet metal
- 1/8″, 1/4″, and 5/16″ drill bits
- Cut-off tool with thin cutting wheel
- Side cutters

Safety Notice: Before performing this job, review all pertinent safety information in the text and discuss safety procedures with your instructor.

Procedure
Practice Beads
Before you begin, make a few continuous welds to get the feel of the gun and to check the adjustments. You will rarely make a continuous weld this long in automotive panel repair because the heat would distort/warp the sheet metal. Take the time to properly adjust the MIG welder. Master the simple welds first. Practice all of these welds until you can make each one perfectly.

1. Secure the hood or sheet metal to a table or welding bench. You will practice welding in the flat position. Completed ❏

2. Make some dry runs with the gun. Do not pull the trigger, just practice maintaining consistent gun angle, travel angle, and travel speed. Completed ❏

3. Connect the work clamp to the hood or sheet metal panel. Completed ❏

4. Check the label inside the welder for recommended settings.
   What is the recommended wire speed setting?

   _____

   What is the recommended voltage setting?

   _____ Completed ❏

5. Set the wire speed and voltage to the recommended values. Completed ❏

Copyright by Goodheart-Willcox Co., Inc. May not be reproduced or posted to a publicly accessible website. **233**

6. Adjust the shielding gas to flow at a rate of 20–25 cfh. Completed ❏

7. Put on your safety gear. Completed ❏

8. Keep the stick-out at a consistent 1/4″. Cut the wire as needed. Completed ❏

9. Pull the trigger and run a 2″ bead using the pull travel direction in the flat position. Completed ❏

10. Watch the arc and listen. You want to hear the steady buzz of a properly set up MIG welder. Completed ❏

Note: If you do not get an arc when you pull the trigger, check to make sure the work clamp is attached to bare metal. See **Figure 3-1**.

Figure 3-1. Before welding, make sure the work clamp is securely attached to bare metal.

11. Adjust the voltage setting slightly up or down to change the size of the bead. Completed ❏

12. Adjust your travel angle, in the range of 90° to 70°, to fine tune penetration. Completed ❏

13. Adjust your travel speed to fine tune penetration. Completed ❏

14. Scribe several 2″ lines. Practice welding *along each line*. Do *not* weld on either side of the line. Try to weld directly on the scribed lines. See **Figure 3-2**. Completed ❏

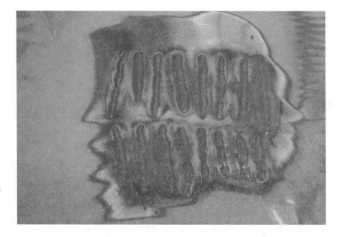

Figure 3-2. The photo shows the workpiece with welds placed over the scribed lines.

Job 3 MIG Welding Basics

Name _____

15. Scribe 10 marks 1/2″ apart and 10 marks 3/4″ apart. Weld a bead between the marks.

    Completed ❑

16. Practice manipulating bead width by varying travel speed. Make a narrow (1/8″ or 3 mm) weld by *increasing* travel speed. Make a wide (3/8″ or 9 mm) weld by *decreasing* travel speed.

    Completed ❑

17. Practice weave travel pattern (moving the gun slightly left and right as you travel) and a spiral travel pattern (making small circles with the gun as you travel). See **Figure 3-3**. These travel patterns decrease travel speed and adjust the travel angle to prevent burn through.

    Completed ❑

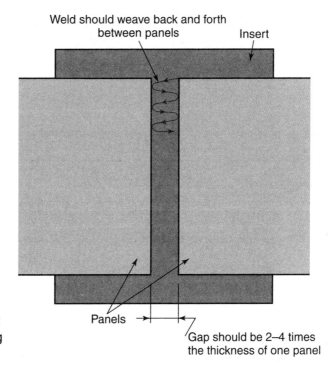

Figure 3-3. The weave technique offers excellent penetration and gap-filling properties.

18. Drill a 5/16″ diameter hole in a 3″ × 3″ square piece of sheet metal. Use this as a template to scribe ten 5/16″ diameter circles in bare metal. Practice welding around the scribed circles. This helps build the skills needed to make plug welds.

    Completed ❑

19. Use a thin cut-off wheel to cut a slit in the hood or the piece of sheet metal. Starting in the middle of the cut, place a spot weld at approximately every 1/2″ along the slit. See **Figure 3-4**. Position the gun at a 90° travel angle to make these spot welds. Completed ❏

Figure 3-4. Place the first spot weld in the middle of the cut as shown. Then place welds every 1/2″ along the cut.

20. Fill the gaps between the spot welds with skip welds to form continuous welds. Completed ❏

21. Make practice beads while holding the gun at a 45° angle. This is the proper gun angle for lap joints. It will allow equal melting of both panels. Completed ❏

22. Drill 1/8″ and 1/4″ holes in the practice hood or a piece of sheet metal. Practice filling each hole by making a spot weld at an edge. Continue making spot welds around the holes until they are filled. Completed ❏

Instructor's Initials _____

Date _____

Name _____ Date _____ Class _____

Job 4

MIG Welding Practice

Objective

After completing this job, you will be able to make spot welds, continuous welds, and plug welds on lap, flange, and butt joints.

Equipment and Materials

To complete this job, you will need the following:
- Four 3″ × 12″ sections of 18–22 gauge sheet metal
- MIG welder
- Cut-off tool with cutting wheel
- Flange tool
- Butt welding clamps
- C-clamps or locking pliers

Safety Notice: Before performing this job, review all pertinent safety information in the text and discuss safety procedures with your instructor.

Procedure

Lap Joint

1. Make a 5/16″ plug weld hole every 2″ along the edge of one of the 3″ × 12″ panels. This is panel #1.   Completed ❏
2. Overlap the plug weld holes onto another panel. This is panel #2.   Completed ❏
3. Clamp the panels together.   Completed ❏

4. Drill 1/8″ holes between the plug weld holes and insert sheet metal screws to hold the panels together. See **Figure 4-1**. Completed ❏

Figure 4-1. Sheet metal screws can be used to hold the panels together before welding.

5. Weld the plug weld holes. Completed ❏
6. Weld the lap joint with skip welds (stitch welds) to make a continuous weld. Completed ❏

Flange Joint

7. Flange the open edge of panel #2. Completed ❏
8. Position a third panel (panel #3) in the flange so there is a 1/4″ gap between the edge of panel #3 and the corner of the flange. Completed ❏
9. Clamp panel #3 in place with a C-clamp or locking pliers. Completed ❏
10. Use a weave or spiral travel pattern to make the skip welds on the flange joint. Completed ❏

Butt Weld

11. Use butt weld clamps to hold the remaining panel (panel #4) to the welded assembly. See **Figure 4-2**. Completed ❏

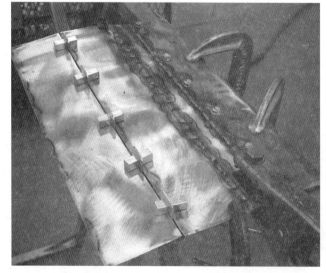

Figure 4-2. Butt weld clamps help maintain the proper panel gap and height during welding.

Chapter 4 MIG Weld Practice

Name _____

12. Make a spot weld near each butt weld clamp.   Completed ❑

13. Remove the butt weld clamps.   Completed ❑

14. Weld the gaps between spot welds with skip welds to make a continuous
    weld.   Completed ❑

Instructor's Initials _____

Date _____

Job 5

Metal Finishing Skills

Objective

After completing this job, you will be able to identify highs and lows by sight and feel, block sand guide coat, and metal finish dents on high-crown panels.

Equipment and Materials

To complete this job, you will need the following:
- Panel with dents
- Shop towel
- Straightedge
- Panel sprayed with surfacer
- Sanding block with 320-grit sandpaper
- Scrap fenders
- Round bar stock
- Pick hammer
- Chisel hammer
- Dolly
- Metal file
- Air grinder

Safety Notice: Before performing this job, review all pertinent safety information in the text and discuss safety procedures with your instructor.

Procedure

Identifying Highs and Lows

1. Use the reflection of the shop's overhead lights in the paint to help locate surface irregularities on the panel surface. Any deviation in the reflection indicates a high or low area. Did you find any irregularities? If so, describe them.

Completed ❏

2. Rub your entire hand (not just your fingers) back and forth over the areas where you visually located irregularities. See **Figure 5-1**. A shop towel or paper towel placed between your hand and the panel will reduce friction and make irregularities easier to feel. Can you feel the irregularities?

_____

Did you feel any surface irregularities that you did not identify in Step 1?

_____ Completed ❑

Figure 5-1. Always use your entire hand when feeling for damage.

3. Place a straightedge on one of the surface irregularities located in the previous step. The straightedge will rock on a high area. A low area will be seen as a gap between the straightedge and the panel. Completed ❑

4. After locating several high and low areas with the straightedge, practice again with your hand until you can easily distinguish between high and low areas. Completed ❑

5. Apply dry guide coat to a panel that already has been sprayed with a coat of surfacer. Completed ❑

6. Block sand the panel with 320-grit sandpaper wrapped around a paint mixing stick. Always sand with the crown. Maintain even pressure on the sanding stick by spreading your fingers evenly along the length of the stick. This technique prevents you from pushing down on one side of the stick more than the other. Stop when you have sanded off almost all the guide coat. Completed ❑

7. After sanding, examine the panel. Any remaining guide coat indicates a low area; bare metal indicates a high area. Continue block sanding until all the guide coat is sanded away or until bare metal is reached in one area. Completed ❑

8. After block sanding, feel the panel. Does the panel feel level? If not, explain.

_____ Completed ❑

Job 5 Metal Finishing Skills   243

Name _____

Unlocking Arrowhead Buckles

9. Obtain a scrap fender. Ask your instructor to make a dent in a high-crown area of a fender that does not have a body line.   Completed ❑

10. Using a dolly with a crown that matches the undamaged crown of the damaged area, tap on the back side of the panel at and just under the tip of the arrowhead. If a second arrowhead is present, tap it in the same manner. Alternately work the two arrowheads.   Completed ❑

11. While holding the dolly firmly against the underside of the arrowhead, use a dinging hammer on the sides of the arrowhead (high areas), starting at the bottom of the arrowhead and working toward the point. Alternate sides of the arrowhead while working toward the point.   Completed ❑

12. After the buckles have been removed, use a hammer-on-dolly technique to raise the shaft of the dent. You may need to hammer hard on the dolly to raise the shaft.   Completed ❑

13. Use a metal file to identify high and low areas in the damaged portion of the panel. High areas are gouged by the file; low areas are untouched by the file and show remaining paint.   Completed ❑

14. Use a pick hammer to raise lows and lower highs.   Completed ❑

15. Continue picking and filing until all lows have been raised and all highs have been lowered.   Completed ❑

Raise Lows and Lower Highs with a Dolly and a Chisel Hammer

16. Obtain a scrap fender. Ask your instructor to make a dent in a high-crown area of a fender that does not have a body line.   Completed ❑

17. Use a dolly with a crown that matches the undamaged crown of the damaged area. Use the hammer-off-dolly technique to release the arrowhead as described in the unlocking arrowhead buckles section (Steps 9–15).   Completed ❑

18. With the arrowheads removed, raise the remaining shaft by placing the flat face of a dolly over the shaft on the outside of the panel. Then use a wide chisel hammer to hammer on the dolly to raise the shaft. Start out with light taps. If the metal will not move, increase the force of the hammer.   Completed ❑

19. When the shaft is raised, make several rapid passes over the damaged area with an air grinder. The grinder should be held flat.   Completed ❑

20. After grinding, examine the panel. Gouged areas indicate highs and painted areas indicate lows. Pick the highs down and pick the lows up.   Completed ❑

21. Continue grinding and picking until the damaged area is level.   Completed ❑

Instructor's Initials _____

Date _____

Name _____ Date _____ Class _____

Job 6

Removing and Installing Bolted Panels

Objective

After completing this job, you will be able to remove and install front and rear bumpers, hoods, fenders, doors, and deck lids.

Equipment and Materials

To complete this job, you will need the following:
- Vehicle with bolted-on doors
- Ratchets, sockets, and wrenches
- Tape measure
- Floor jack

Safety Notice: Before performing this job, review all pertinent safety information in the text and discuss safety procedures with your instructor.

Procedure

Removing Bolted Panels

1. What must be done if hidden damage is found when a bolted panel is removed?

   _____
   _____
   _____
   _____

   Completed ❑

245

Front Bumper Assembly

2. Check the front bumper assembly for fit. Check the gaps or alignment with adjacent body parts. List any problems found.

   _____ Completed ❑

3. List any parts that must be removed for access to the bumper fasteners.

   _____

   _____ Completed ❑

4. Remove the front bumper assembly. Completed ❑

5. To what parts were the bumper fasteners attached?

   _____ Completed ❑

6. What size (in mm) were the bumper fastener bolt heads?

   _____ Completed ❑

7. What types of fasteners hold the bumper assembly together?

   _____ Completed ❑

8. Place all fasteners into a labeled plastic bag. See **Figure 6-1**. Completed ❑

Figure 6-1. Do not lose the fasteners after disassembly. Put them in a labeled container and set it in a secure place.

9. List the components of the bumper assembly here.

   _____

   _____

   _____ Completed ❑

Job 6 Removing and Installing Bolted Panels

Name _____

Hood Assembly

10. Check the hood assembly for proper alignment. Check for a uniform gap at the rear of the hood (cowl vent panel or windshield). Check for uniform gaps at the fenders. Check for a uniform gap at the hood/grille and front bumper assembly or header panel. Write the width of each of the gaps here. Also note any problems with gaps.

    Hood to cowl vent panel or windshield gap: _____

    Hood-to-fender gap: _____

    Left side: _____

    Right side: _____

    Hood to bumper cover, grille, or header panel gap: _____

    Problems found: _____ Completed ❑

11. Mark the location of the hood hinges on the hood. Completed ❑

12. List any parts that must be removed before the hood can be removed.

    _____
    _____ Completed ❑

13. Remove the hood-to-hinge bolts as two students hold the sides of the hood. Set the hood aside. Completed ❑

14. What size (in mm) are the hood-to-hinge bolt heads?

    _____ Completed ❑

15. Place all the fasteners into a labeled plastic bag. Completed ❑

16. List the components of the hood assembly here.

    _____
    _____
    _____
    _____ Completed ❑

Fender Assembly

17. Check for a uniform gap between the fender and the door. Write down the width of the gap and note any problems.

    Fender/door gap: _____ Completed ❑

18. List the parts that must be removed to access the fender fasteners or to remove the fender.

    _____
    _____ Completed ❑

19. Remove the fender fasteners and remove the fender. Completed ❑

20. What type of fasteners hold the fender assembly together?

    _____ Completed ❑

21. What size (in mm) were the fender fastener bolt heads?

    _____ Completed ❑

22. Place all fasteners into a labeled plastic bag. Completed ❑

23. List the components of the fender assembly here.

    _____
    _____ Completed ❑

Door Assembly

24. Check the front door for alignment. First check for a uniform gap between the door and rocker panel; then check the gap between the door window frame and windshield pillar. Finally, check the gap between the door and rear door or quarter panel. Write down the width of the gaps below. Note any problems.

    Door/rocker panel gap: _____

    Door/windshield pillar gap: _____

    Door/rear door or quarter panel gap: _____

    Problems found: _____ Completed ❑

25. If the vehicle has electrical wires entering the door, look for an electrical junction between the door and vehicle. If there is no electrical junction, the door panel must be removed and the individual wires disconnected from the electrical components inside the door. The wires are then pulled out of the door. Completed ❑

26. List the components, including wiring, that must be removed before the door can be removed.

    _____
    _____ Completed ❑

27. Mark the location of the door hinges on the door. Support the rear of the door with a floor jack. Position a student to steady the door on the floor jack. Remove the fasteners that hold the hinges to the door. Leave the hinges on the vehicle. Remove the door and set it aside. Completed ❑

28. What types of fasteners hold the door assembly together?

    _____ Completed ❑

29. What size (in mm) were the door fastener bolt heads?

    _____ Completed ❑

30. Place all fasteners into a labeled plastic bag. Completed ❑

Job 6 Removing and Installing Bolted Panels 249

Name _____

31. List the components of the door assembly here.

_____

_____ Completed ❑

Deck Lid Assembly

32. Check the gaps at the front of the deck lid, between the deck lid and quarter panel, and between the deck lid and taillights. Write down the width of the gaps and note any problems below.

   Front of deck lid gap: _____

   Deck lid/quarter panel gap:

   Left: _____

   Right: _____

   Deck lid/taillights: _____

   Problems found: _____ Completed ❑

33. List any parts that must be removed before the deck lid can be removed.

_____

_____ Completed ❑

34. Mark the location of the deck lid hinges on the deck lid. See **Figure 6-2**. Completed ❑

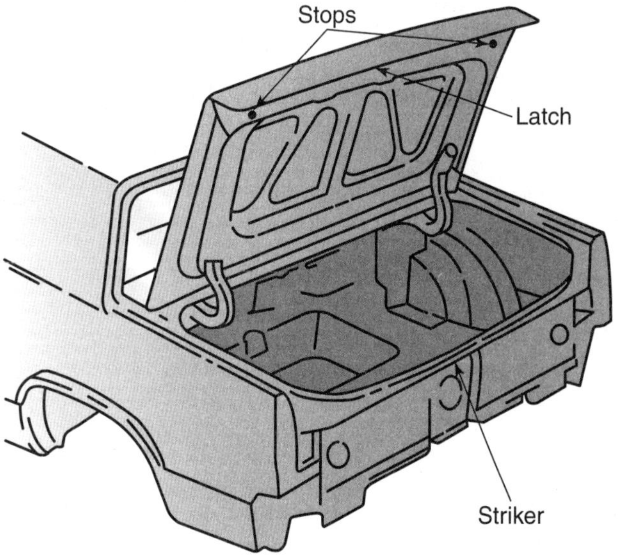

Figure 6-2. Before you remove the deck lid, mark the location of the hinges on the underside of the lid.

35. Position a student on either side of the deck lid and remove the deck lid-to-hinge bolts. Remove the deck lid and set it aside. Completed ❑

36. What size (in mm) were the deck lid bolts?

_____ Completed ❑

Copyright by Goodheart-Willcox Co., Inc. May not be reproduced or posted to a publicly accessible website.

37. What other fasteners hold the deck lid assembly together?

_____ Completed ❑

38. Place all fasteners into a labeled plastic bag. Completed ❑

39. List the components of the deck lid assembly here.

_____

_____

_____ Completed ❑

Rear Bumper Assembly

40. Check the rear bumper assembly for fit. Check the gaps or alignment with adjacent body parts. Completed ❑

41. List any parts that must be removed for access to the bumper fasteners.

_____

_____ Completed ❑

42. Remove the rear bumper assembly. Completed ❑

43. What types of fasteners are used to secure the bumper to the vehicle?

_____ Completed ❑

44. What size (in mm) were the bumper fastener bolt heads?

_____ Completed ❑

45. What types of fasteners hold the bumper assembly together?

_____ Completed ❑

46. Place all fasteners into a labeled plastic bag. Completed ❑

47. List the components of the bumper assembly here.

_____

_____ Completed ❑

Panel Installation

48. Install the rear bumper assembly. Match the original alignment. Completed ❑

49. Install the deck lid. Ask two other students to line up the deck lid hinges with the marks on the deck lid. Tighten the bolts that secure the deck lid to the hinges. Check the panel gaps and the opening and closing of the deck lid. Completed ❑

50. Install the door. Support the door on a floor jack and move the door into position. Install the bolts and align the hinges with the marks on the door. Check for a uniform gap between the door and the rocker panel. Next, check the gap between the door and windshield pillar, and then check the gap between the door and the quarter panel or rear door. Check for proper opening and closing. Completed ❑

Job 6　Removing and Installing Bolted Panels　　　251

Name _____

51. Install the fender on the vehicle. Install the fender bolts but do not tighten them. Check for a uniform gap between the fender and door.　　Completed ❑

52. Install the hood. With the help of two other students, line up the marks on the hood with the hinges. Install and tighten the bolts. Check for gaps between the hood and the cowl vent panel/windshield, fender and header panel/front bumper. If the fender/hood gap is good, tighten the fender bolts. Check for proper opening and closing.　　Completed ❑

53. Install the front bumper. Match the original alignment.　　Completed ❑

Instructor's Initials _____

Date _____

Name _____ Date _____ Class _____

Job 7

Panel Splicing

Objective

After completing this job, you will be able to splice panels using butt welds.

Equipment and Materials

To complete this job, you will need the following:
- MIG welder
- Scrap fender
- Cut-off tool with cutting and grinding wheels
- C-clamps, locking pliers, or butt-weld clamps
- Masking tape

Safety Notice: Before performing this job, review all pertinent safety information in the text and discuss safety procedures with your instructor.

Procedure

Splicing a Fender

1. Remove the paint from a 6″-wide vertical area between the wheel opening and the top of the fender.   Completed ❑
2. Mark a cut line with masking tape.   Completed ❑
3. Use a cut-off wheel to cut along the edge of the tape.   Completed ❑
4. Clamp along the edges of the cut in preparation for a butt weld. Use butt weld clamps if available. If butt weld clamps are not available, clamp the edges together so that a gap the thickness of the metal remains between the panel edges. Use enough clamps to prevent the panel edges from moving during welding.   Completed ❑
5. Make spot welds between the clamps.   Completed ❑
6. Remove clamps.   Completed ❑
7. Complete the butt joint using stitch welds to form a continuous weld.   Completed ❑

Note: Minimize heat distortion by making a small stitch weld and allowing the panel to cool before making the next weld. Never try to hurry the process by cooling a hot weld with water, as this will make the metal stiff and brittle.

8. Grind the welds with a cut-off tool equipped with a grinding wheel. Grinding heat can easily warp thin automotive sheet metal. Grind for 10 seconds and then allow the weld to cool for 30 seconds. Grind on the welds only; do not grind on the surrounding metal. See **Figure 7-1**.

Completed ❏

Figure 7-1. Grind only on the protruding weld. Also, take frequent breaks to let the weld cool. Grinding can generate enough heat to warp automotive sheet metal.

Instructor's Initials _____

Date _____

Name _____ Date _____ Class _____

Job 8

Flexible Plastic Repair

Objective

After completing this job, you will be able to clean a flexible plastic panel, apply plastic repair material, and finish sand-cured plastic repair material.

Equipment and Materials

To complete this job, you will need the following:
- Plastic bumper cover (removed from vehicle)
- Dual-action (DA) sander
- Cut-off tool
- Aluminum body tape
- Adhesion promoter
- Flexible plastic repair material
- Fiberglass cloth
- 80-, 180-, and 320-grit sandpaper
- Plastic cleaner
- Soap and water

Safety Notice: Before performing this job, review all pertinent safety information in the text and discuss safety procedures with your instructor.

Procedure
Cleaning

1. Clean the inside and outside of the plastic bumper cover with soap and water. Then clean it again using plastic cleaner. Completed ❏

2. Identify plastic type by locating the identification label (ISO code). What type of plastic is the bumper cover made from?

   _____

   _____ Completed ❏

3. Use a cut-off tool to make a 2″ cut through the plastic in a low-crown part of the bumper cover. Completed ❏

4. Use a cut-off tool to make a 2″ gouge in a high-crown part of the bumper cover. When making the gouge, do not cut through the plastic. Completed ❑

Featheredging Damaged Area

5. Sand both sides of the cut to a taper with 80-grit sandpaper. This taper should be deepest at the cut and gradually become shallower. The deepest point of the taper should be halfway through the bumper cover. Completed ❑

Note: A DA sander can be used when tapering non-olefin plastic. See **Figure 8-1**. Olefin bumper covers should be hand sanded, as a DA sander can generate enough heat to melt the plastic.

Figure 8-1. Bevel both sides of the damage with 80-grit on a DA.

Bevel

6. Featheredge the taper and surrounding paint with 180-grit sandpaper. Complete the featheredge of the exposed plastic and surrounding paint with 320-grit sandpaper. The featheredge of existing paint should extend at least 2″ around the cut. Completed ❑

7. Taper the sides of the gouge with 80-grit sandpaper. The depth of the taper should be half the thickness of the bumper cover. Completed ❑

8. Featheredge the taper and surrounding paint with 180-grit sandpaper. Complete the featheredge of the exposed plastic and surrounding paint with 320-grit sandpaper. The featheredge of existing paint should extend at least 2″ around the gouge. Completed ❑

Reinforcing the Damaged Area

9. Scuff the back side of the cut with 80-grit sandpaper. The scuffing should extend at least 2″ around the cut. Completed ❑

10. Clean off the sanding dust by blowing off the repair area. Completed ❑

11. Align the cut edges and apply aluminum tape to the front side of the cut to keep the edges in alignment. Completed ❑

12. Read the instructions on the container of plastic repair material. Completed ❑

Job 8 Flexible Plastic Repair

Name _____

13. Summarize the instructions here.

   _____
   _____
   _____
   _____   Completed ❑

14. Is adhesion promoter required?

   _____   Completed ❑

15. If adhesion promoter is required, apply it to the scuffed area on the back side of the cut.   Completed ❑

16. Follow the mixing instructions and mix enough flexible repair material to cover the scuffed area on the back side of the cut.   Completed ❑

17. Apply the mixed flexible repair material to the back side of the cut within the scuffed area.   Completed ❑

18. Cut out a 4" × 2" section of fiberglass cloth to serve as reinforcement.   Completed ❑

19. Apply the 4" × 2" fiberglass cloth reinforcement to the flexible repair material on the back side of the cut. See **Figure 8-2**.   Completed ❑

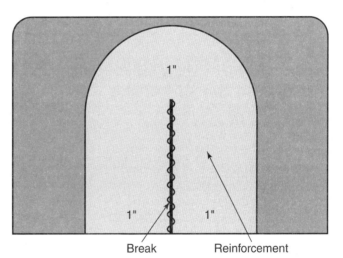

Figure 8-2. Apply the 4" × 2" fiberglass cloth reinforcement to the flexible repair material on the back side of the damage.

20. Allow the flexible repair material to cure.   Completed ❑

Filling with Repair Material

21. Remove the tape from the cut.   Completed ❑

22. If an adhesion promoter is required, apply adhesion promoter to the tapered and featheredged areas on the front of the cut and the gouge.   Completed ❑

23. Does the plastic repair material manufacturer recommend applying the repair material over the paint featheredge? Follow the manufacturer's recommendations.

   _____   Completed ❏

24. Mix enough flexible repair material to completely fill the cut and the gouge.   Completed ❏

25. Apply an overfill of flexible repair material to the cut and gouge.   Completed ❏

26. Allow the flexible plastic repair material to cure.   Completed ❏

Sanding the Repair Area

27. Sand the cured plastic repair material with 80-grit paper followed with 180-grit paper. Use the edge of the sanding stick when sanding the high-crown area. Use a flexible pad to sand double-crown areas.   Completed ❏

28. Inspect the sanded material for pinholes or undercut. Fill pinholes and undercut areas with additional plastic repair material.   Completed ❏

29. Allow the reapplication of flexible repair material to cure.   Completed ❏

30. Finish sand the cured repair material to the proper contour with 180-grit sandpaper.   Completed ❏

Instructor's Initials _____

Date _____

Name _____ Date _____ Class _____

Job 9

Measurements

Objective

After completing this job, you will be able to select measuring points on a vehicle and measure the distance between these points with a tape measure and a tram.

Equipment and Materials

To complete this job, you will need the following:
- Tape measure graduated in inch and millimeter increments
- Tram graduated in inch and millimeter increments
- Vehicle

Safety Notice: Before performing this job, review all pertinent safety information in the text and discuss safety procedures with your instructor.

Procedure

Taking Tape Measurements

Note: All measurements required in the following section should be taken using a tape measure with inch and millimeter increments. Record your measurements in both inches and millimeters.

Hood Opening

1. Measure the center-to-center distance between the front and rear front fender bolts on the left side of the vehicle. Record your measurement.

    Center-to-center distance: _____ in., _____ mm   Completed ❑

2. Measure the center-to-center distance between the front and rear front fender bolts on the right side of the vehicle. Record your measurement.

    Center-to-center distance: _____ in., _____ mm   Completed ❑

3. Did you need to bend the tape measure to make these measurements?

    _____
                                                                          Completed ❑

259

4. Perform an X-check of the hood opening by measuring the center-to-center distance between the front and rear fender bolts. Record your measurements.

   Left front to right rear: _____ in., _____ mm

   Right front to left rear: _____ in., _____ mm

   Completed ❑

5. Did you need to bend the tape to make these measurements?

   _____

   Completed ❑

Door Opening

6. Measure from the upper front of the door opening to the lower rear of the door opening. Record your measurement.

   Upper front to lower rear: _____ in., _____ mm

   Completed ❑

7. Measure from the upper rear of the door opening to lower front of the door opening. Record your measurement.

   Upper rear to lower front: _____ in., _____ mm

   Completed ❑

8. Measure the distance from the door striker to the front of the door opening. Record your measurement.

   Striker to front of opening: _____ in., _____ mm

   Completed ❑

9. Did you have to bend the tape to make the measurements?

   _____

   Completed ❑

Deck Lid Opening

10. Remove the weather strip and find seams that can be used as reference points.

    Completed ❑

11. Make an X-check measurement of the deck lid opening. Record your measurements.

    Left front to right rear: _____ in., _____ mm

    Right front to left rear: _____ in., _____ mm

    Completed ❑

12. Did you have to bend the tape to make any of the measurements?

    _____

    Completed ❑

Wheelbase

13. With the front wheels pointed straight ahead, measure the distance between the center of the front wheel and the center of the rear wheel on each side of the vehicle. Record your measurements.

    Left side: _____ in., _____ mm

    Right side: _____ in., _____ mm

    Completed ❑

14. Did you have to bend the tape to make these measurements?

    _____

    Completed ❑

Job 9 Measurements

Name _____

Toe

15. With the front wheels pointed straight ahead, measure the center-to-center distance at the front and rear of the front tires. Record your measurements.

    Front of tires _____ in., _____ mm

    Rear of tires _____ in., _____ mm

    Completed ❏

16. Did you have to bend the tape to make any of the measurements?
    _____

    Completed ❏

17. What happens to the accuracy of any measurement when the tape is bent to make a measurement?
    _____

    Completed ❏

Taking Tram Measurements

Note: All measurements in the following section should be taken using a tram graduated in inch and millimeter increments.

Hood Opening

18. How long are the pointers on the tram?

    _____ in., _____ mm

    Completed ❏

19. Measure the center-to-center distance between the front and rear front fender bolts on the left side of the vehicle. Record your measurement.

    Center-to-center distance: _____ in., _____ mm

    Completed ❏

20. Measure the center-to-center distance between the front and rear front fender bolts on the right side of the vehicle. Record your measurement.

    Center-to-center distance: _____ in., _____ mm

    Completed ❏

21. Measure the center-to-center distance between the front bolts on the right and left front fenders of the vehicle. Record your measurement.

    Center-to-center distance: _____ in., _____ mm

    Completed ❏

22. Measure the center-to-center distance between rear bolts on the right and left front fenders of the vehicle. Record your measurement.

    Center-to-center distance: _____ in., _____ mm

    Completed ❏

23. Perform an X-check of the hood opening using the center-to-center distance between the front and rear bolts as reference points. See **Figure 9-1**. Record your measurement.

    Left front to right rear: _____ in., _____ mm

    Right front to left rear: _____ in., _____ mm   Completed ❑

Figure 9-1. This technician is using a tram to perform an X-check. An X-check is an excellent way to tell if the opening is out-of-square.

24. Was the tram able to reach around all obstructions?

    _____   Completed ❑

25. Did the measuring arms get in the way?

    _____   Completed ❑

Door Opening

26. Measure from the upper front of the door opening to the lower rear of the door opening. Record your measurement.

    Upper front to lower rear: _____ in., _____ mm   Completed ❑

27. Measure from the upper rear of the door opening to the lower front of the opening. Record your measurement.

    Upper rear to lower front: _____ in., _____ mm   Completed ❑

28. Measure the distance from the door striker to the front of the door opening. Record your measurement.

    Striker to front of opening: _____ in., _____ mm   Completed ❑

29. Was the tram able to reach around all obstructions?

    _____   Completed ❑

30. Did the measuring arms get in the way?

    _____   Completed ❑

Job 9 Measurements

Name _____

Deck Lid Opening

31. Remove the weather strip and find seams to use as reference points. Completed ❏

32. Make an X-check measurement of the deck lid opening. Record your measurements below.

    Left front to right rear: _____ in., _____ mm

    Right front to left rear: _____ in., _____ mm Completed ❏

33. Was the tram able to reach around all obstructions?
    _____ Completed ❏

34. Did the measuring arms get in the way?
    _____ Completed ❏

Wheelbase

35. With the front wheels pointed straight ahead, measure the center-to-center distance between the front and rear wheels on both sides of the vehicle. Record your measurements below.

    Left side: _____ in., _____ mm

    Right side: _____ in., _____ mm Completed ❏

36. Was the tram able to reach around all obstructions?
    _____ Completed ❏

37. Did the measuring arms get in the way?
    _____ Completed ❏

Toe

38. With the front wheels pointed straight ahead, measure the center-to-center distance between the front and rear of the front tires. Record your measurements below.

    Front of tires: _____ in., _____ mm

    Rear of tires: _____ in., _____ mm Completed ❏

39. Was the tram able to reach around all obstructions?
    _____ Completed ❏

40. Did the measuring arms get in the way?
    _____ Completed ❏

41. Compare the use of a tram with a tape measure. In which situations is a tram more accurate than a tape measure? When is a tape measure more accurate than a tram?

_____

_____

_____

_____ Completed ❑

Instructor's Initials _____

Date _____

Name _____ Date _____ Class _____

Job 10

Removing and Installing Short Arm/Long Arm Suspension Components

Objective

After completing this job, you will be able to test for ball joint wear; measure ball joint location; repack and adjust wheel bearings; measure brake pad thickness; and remove and install brake calipers, steering knuckles, and control arms.

Equipment and Materials

To complete this job, you will need the following:
- Lift or a floor jack and jack stands
- Wheel chocks
- Tape measure
- Brake cleaner
- Wheel bearing grease
- Wheel bearing grease seal
- Coil spring compressor
- Cotter pin
- Torque wrench
- Hammer
- Thin, flat-blade screwdriver
- Ratchet, sockets, and wrenches
- Vehicle with a short arm/long arm suspension

Safety Notice: Before performing this job, review all pertinent safety information in the text and discuss safety procedures with your instructor.

Procedure

1. Turn the steering wheel so the front wheels are pointing straight ahead. Safely raise the vehicle with a lift. If a lift is not available, chock the rear wheels, raise the vehicle with a floor jack, and place the vehicle on jack stands.   Completed ❏

2. While holding the front tire with your hands placed at 12 o'clock and six o'clock, try to rock the wheel by pushing in at the top and pulling out at the bottom. Then pull out at the top and push in at the bottom. Play indicates a worn ball joint. Test both front wheels.   Completed ❏

3. Remove the front wheels. Completed ❑

4. Locate the following parts of the long arm/short arm suspension system: upper control arm, lower control arm, upper ball joint, lower ball joint, steering knuckle, coil spring, shock absorber, brake caliper, and brake rotor/hub. Completed ❑

5. Obtain a service manual for the vehicle you are working on. Read the service manual to find the torque values for the fasteners used to secure the components listed below. List each fastener type and its torque value in the space provided. Identify any single-use fasteners with a *.

   Brake caliper: _____

   Brake rotor/hub: _____

   Shock absorber: _____

   Coil spring: _____

   Upper ball joint: _____

   Lower ball joint: _____

   Steering knuckle: _____

   Upper control arm: _____

   Lower control arm: _____ Completed ❑

6. Find symmetrical reference points on the frame for measuring ball joint location. Use a tape measure to measure from the center of the lower ball joint to a reference point on the frame on both the left and right sides of the vehicle.

   Left ball joint measurement _____ in., _____ mm

   Right ball joint measurement _____ in., _____ mm

   This measurement is a quick check for suspension damage. An impact on a front wheel can bend the control arms and steering knuckle, moving the ball joint toward the rear of the vehicle. Completed ❑

7. As fasteners are removed, clean each fastener with solvent and wipe the fasteners off with a rag. Inspect the threads of each fastener for damage or rust. Use a tap or a die to chase the threads. Check the hex portion of the fastener for rounding. Replace any rounded fasteners. Place fasteners into a labeled container. Completed ❑

8. Remove the brake caliper. Use mechanic's wire or piece of MIG welding wire to hang the caliper from the frame or unibody. Do not allow the caliper to hang by the brake hose, as this can damage the hose. Completed ❑

9. Remove the grease cap by carefully prying with a thin, flat-blade screwdriver. Straighten the cotter pin with pliers and pull it from the nut lock or adjusting nut. Remove the nut lock (if used), adjusting nut, and thrust washer. Clean your hands. Carefully slide the brake rotor/hub off the spindle portion of the steering knuckle. Catch the outer wheel bearing in a clean hand. Place a clean rag inside the hub to keep out grit. Store the rotor/hub in a clean area. Completed ❑

Job 10 Removing and Installing Short Arm/Long Arm Suspension Components

Name _____

10. Remove the inner wheel bearing. Place the wheel bearings into a sealable plastic bag and seal the bag to keep out grit. — Completed ❑

11. Remove the upper and lower fasteners and remove the shock absorber. — Completed ❑

12. Remove the cotter pin and nut from the tie rod end. Tap on the steering arm with a hammer until the tie rod end loosens. Remove the tie rod end. — Completed ❑

13. Install a coil spring compressor inside the coil spring. Tighten the coil spring compressor to compress the coil spring. — Completed ❑

14. Place a floor jack under the lower control arm. Position the jack so the saddle of the jack does not block access to the lower ball joint. — Completed ❑

15. Straighten and remove the cotter pin from the lower ball joint. Loosen the ball joint nut. — Completed ❑

16. Separate the ball joint from the steering knuckle with a two-prong fork tool or by tapping on the steering knuckle with a large hammer. — Completed ❑

17. Remove the ball joint nut. — Completed ❑

18. Lower the floor jack to drop the control arm and ball joint away from the steering knuckle. — Completed ❑

19. Remove the coil spring and spring insulators. Slowly release the spring from the compressor. Inspect the spring insulators for wear and cracks. Obtain replacement insulators for worn or cracked insulators. — Completed ❑

20. Straighten and remove the upper ball joint cotter pin. Loosen the upper ball joint nut. — Completed ❑

21. Strike the steering knuckle with a hammer to separate the upper ball joint. — Completed ❑

22. Remove the upper ball joint nut. — Completed ❑

23. Remove the steering knuckle. — Completed ❑

24. Remove the upper control arm fasteners and remove the upper control arm. — Completed ❑

25. Remove the lower control arm fasteners and remove the lower control arm. — Completed ❑

26. Inspect the control arm bushings for hardening, damage, wear, and cracks. Obtain replacement bushings for worn or cracked bushings. — Completed ❑

27. If a ball joint is loose, replace it. — Completed ❑

28. Install the control arm bushings, upper and lower control arms, and fasteners. Position the floor jack under the lower control arm. Tighten fasteners to the specified torque value. — Completed ❑

29. Install the coil spring insulators on the control arms. — Completed ❑

30. Compress the spring and insert it into position between the control arms. See **Figure 10-1**. Completed ❑

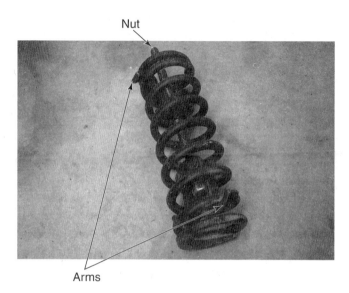

Figure 10-1. A spring compressor is often used when removing and installing a coil spring. This compressor has two arms that attach to the spring's coils. Tightening a nut on the threaded rod brings the arms together, compressing the spring.

31. Raise the lower control arm with the floor jack. Completed ❑

32. Adjust the height of the lower control arm with the floor jack to properly position the control arms so the steering knuckle will fit. Install the steering knuckle and nuts on the ball joints. Tighten the fasteners to the specified torque value. Install new cotter pins. If the hole in the ball joint stud does not line up with a hole in the ball joint nut, tighten the nut slightly to align the holes and allow the cotter pin to slip through all holes. Bend the ends of the cotter pin. What is the proper way to bend the ends of a cotter pin?

_____

_____ Completed ❑

33. Lower the floor jack. Completed ❑

34. Remove the coil spring compressor. Completed ❑

35. Install the tie rod end. Torque the nut to the specified value. Install the cotter pin. Completed ❑

36. Remove the grease seal from the rotor/hub. Clean the inside of the rotor/hub and the inner and outer wheel bearings with solvent. Allow the parts to air dry. Do not use compressed air to speed the drying process. Every trace of old grease must be removed to prevent contamination. Completed ❑

37. Inspect the wheel bearings for wear, damage, and discoloration. Replace the wheel bearing and races if any wear, damage, or discoloration is found. Completed ❑

38. Place wheel bearing grease inside the hub. Completed ❑

Job 10 Removing and Installing Short Arm/Long Arm Suspension Components

Name _____

39. Repack the wheel bearing with wheel bearing grease. Put on a plastic glove. Place the wheel bearing, small diameter side down, into the gloved hand. Push/force wheel bearing grease into the large diameter side of the wheel bearing and thoroughly coat the rollers and the cages. Rotate the cages to check for complete grease penetration. Completed ❑

40. Clean the spindle with solvent. Then smear the spindle with wheel bearing grease. Completed ❑

41. Install the inner wheel bearing in the hub. Completed ❑

42. Install the new wheel bearing grease seal on the hub. Completed ❑

43. Install the rotor/hub on the spindle. Completed ❑

44. Install the outer wheel bearing into the hub. Completed ❑

45. Carefully rotate the rotor/hub to set the bearings in place. Completed ❑

46. Install the thrust washer and adjusting nut. Completed ❑

47. Rotate the rotor/hub ten turns. Use a torque wrench to tighten the adjusting nut to 15 ft lb. Then loosen the adjusting nut and retighten it to 1 ft lb. (10–15 in lb). Completed ❑

48. Install the nut lock (if used). If the new cotter pin will not align with the holes in the spindle, slightly loosen the adjusting nut to allow alignment. Install the new cotter pin and bend the ends. Completed ❑

49. Inspect the caliper and caliper mounts on the steering knuckle for damage. Measure the thickness of the brake pads.

    Inner brake pad thickness: _____ in., _____ mm

    Outer brake pad thickness: _____ in., _____ mm

    Do the brake pads need to be replaced? _____ Completed ❑

50. Clean the inner and outer surfaces of the rotor with brake cleaner. Completed ❑

51. Install the caliper on the steering knuckle mounts. Tighten the fasteners to the specified torque value. Completed ❑

52. Install the front wheel and torque the wheel lug nuts to the specified value. Completed ❑

53. Measure ball joint location:

    Left ball joint measurement: _____ in., _____ mm

    Right ball joint measurement: _____ in., _____ mm Completed ❑

54. Lower the vehicle from the lift or jack stands. Completed ❑

Instructor's Initials _____

Date _____

Name _____ Date _____ Class _____

Job 11

Removing and Installing a MacPherson Strut Assembly

Objective

After completing this job, you will be able to remove and install a MacPherson strut, lower control arm, and steering knuckle as an assembly.

Equipment and Materials

To complete this job, you will need the following:
- Vehicle equipped with a MacPherson strut suspension
- Ratchets, sockets, and wrenches
- Lift or jack stands, wheel chocks, and a floor jack
- Brake cleaner
- Tape measure

Safety Notice: Before performing this job, review all pertinent safety information in the text and discuss safety procedures with your instructor.

Procedure

1. Turn the steering wheel so the front wheels are pointing straight ahead.   Completed ❏

2. Safely raise the vehicle with a lift. If a lift is not available, chock the rear wheels, raise the front of the vehicle with a floor jack, and lower the front of the vehicle onto jack stands.   Completed ❏

3. Remove the front wheels.   Completed ❏

4. Locate the drive axle nut. Check the nut to see if there is an indentation (stake) in the nut at the threads. If a stake is present, use a punch to straighten the indentation.   Completed ❏

5. Loosen the drive axle nut.   Completed ❏

6. Locate the following parts of the MacPherson strut assembly: stabilizer bar, stabilizer bar bushings, stabilizer bar links, lower control arm, lower ball joint, steering knuckle, coil spring, and strut.   Completed ❏

Copyright by Goodheart-Willcox Co., Inc. May not be reproduced or posted to a publicly accessible website.

271

7. Obtain a service manual for the vehicle you are working on. Locate the torque values for all the fasteners used to secure the components listed below. List each fastener type and its torque value in the space provided. If there are multiple fasteners and torque values, be sure to list them all. Identify any single-use fasteners with a *.

   Wheel: _____

   Stabilizer bar: _____

   Stabilizer bar link: _____

   Brake caliper: _____

   Drive axle nut: _____

   Tie rod end: _____

   Strut nuts in shock tower: _____

   Lower control arm to unibody: _____ Completed ❏

8. Find symmetrical reference points on the frame for measuring ball joint location. Use a tape measure to measure from the center of the lower ball joint to a reference point on the frame on both the left and right sides of the vehicle. Record the results below. Be sure to indicate units of measure.

   Lower left ball joint measurement _____ in., _____ mm

   Lower right ball joint measurement _____ in., _____ mm

   These measurements can be used as a quick check for suspension damage. An impact on a front wheel can bend control arms and steering knuckle, moving the ball joint toward the rear of the vehicle. Completed ❏

9. Clean each removed fastener with solvent. Wipe the fastener off with a rag. Inspect the threads of the fastener for damage or rust. Use a die to chase the threads. Check the hex portion of the fastener for rounding. Replace any rounded fasteners. Place fasteners in a labeled container. Completed ❏

10. Remove brake caliper. Use mechanic's wire or a piece of MIG welding wire to hang the caliper from the unibody. Do not allow the caliper to hang by the brake hose. Hanging the caliper by the brake hose will damage the hose. Completed ❏

11. Remove the brake rotor. Completed ❏

12. Count the number of threads visible between the tie rod end and the adjusting sleeve. Write the number of threads in the space provided.

    _____ Completed ❏

13. Straighten and remove the cotter pin from the tie rod end nut. Completed ❏

14. Remove the tie rod end nut. Completed ❏

15. Tap on the steering knuckle with a large hammer to break the tie rod end loose from the knuckle. Completed ❏

16. Remove the tie rod end from the steering knuckle. Completed ❏

17. Unbolt the stabilizer bar link from the lower control arm. Completed ❏

Job 11 Removing and Installing a MacPherson Strut Assembly

Name _____

18. Position a floor jack under the lower control arm to support the strut, control arm, and steering knuckle assembly.   Completed ❑

19. Unbolt the lower control arm from the unibody.   Completed ❑

20. Turn the drive axle nut so the nut is at the end of the shaft. Hit the nut with a hammer to break the drive axle loose from the steering knuckle.   Completed ❑

21. Remove the drive axle nut. Pull the suspension assembly away from the vehicle while pushing the shaft toward the center of the vehicle to separate the steering knuckle and drive axle.   Completed ❑

22. Do not allow the drive axle to hang by the inner CV joint. Use wire to suspend the outer CV joint from the body.   Completed ❑

23. Unbolt the MacPherson strut from the shock tower. See **Figure 11-1**.   Completed ❑

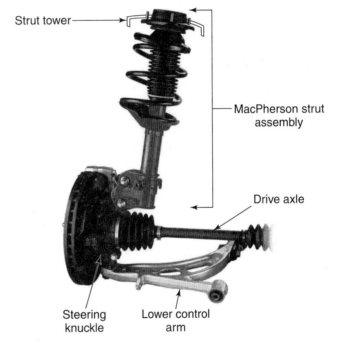

Figure 11-1. A MacPherson strut assembly is a self-contained unit that consists of a shock absorber and a coil spring.

24. Roll the strut, control arm, and steering knuckle assembly out from under the vehicle with the floor jack.   Completed ❑

25. To install the strut, control arm, and steering knuckle assembly, roll the assembly back into position under the vehicle.   Completed ❑

26. Fit the top of the strut assembly into the strut tower. Install the nuts needed to hold the strut to the strut tower. Tighten the fasteners to the specified torque value.   Completed ❑

27. Align the splines to fit the steering knuckle over the outer CV joint shaft. Install the axle nut. Attach the lower control arm to the unibody. Tighten the fasteners to the specified torque value.   Completed ❑

28. Check to see that the original number of threads are visible at the tie rod adjusting sleeve. If the number of threads is not correct, turn the tie rod end until the number of threads visible is correct. Install the tie rod in the steering knuckle and tighten the nut to the specified torque value. Install a new cotter pin. If the cotter pin will not align with the holes, tighten the nut slightly to align the holes. If necessary, connect the speed sensor wire. Completed ❑

29. Install the stabilizer bar links. Tighten all fasteners to the specified torque value. Completed ❑

30. Clean the brake rotor with brake cleaner. Install the brake rotor. Completed ❑

31. Bolt the brake caliper in place and torque the fasteners to the specified value. Completed ❑

32. Use a tape measure to measure from the center of the lower ball joint to a reference point on the frame on both the left and right sides of the vehicle. Record the results below. Be sure to indicate units of measure.

    Lower left ball joint measurement _____ in., _____ mm

    Lower right ball joint measurement _____ in., _____ mm   Completed ❑

33. Install the front wheels. Torque the wheel lug nuts to the specified value. Completed ❑

34. Lower the vehicle from the lift or jack stands. Completed ❑

35. Tighten the axle nut and the wheel lug nuts to the specified torque value. Completed ❑

36. Install the hub cap if necessary. Completed ❑

Instructor's Initials _____

Date _____

Name _____ Date _____ Class _____

Job 12

Removing and Installing a Radiator

Objective

After completing this job, you will be able to remove and install a radiator in a front-wheel-drive vehicle.

Equipment and Materials

To complete this job, you will need the following:
- Front-wheel-drive vehicle with an electric radiator fan
- Drain pans
- Ratchet, sockets, and wrenches

Safety Notice: Before performing this job, review all pertinent safety information in the text and discuss safety procedures with your instructor.

Procedure

Radiator Removal

1. Disconnect the negative battery cable and tie it down so it cannot accidently touch the negative battery terminal. Completed ❏

2. If necessary, allow the engine to cool. Completed ❏

2a. Position a drain pan under the vehicle. Search the radiator for a drain valve. If you find a drain valve, open it. Drain the radiator into the drain pan. If a valve cannot be found, remove the lower radiator hose by loosening the hose clamp at the radiator. Drain the radiator's contents into the drain pan. Completed ❏

3. Unplug the electric fan motor. If the wiring is attached to the fan shroud, disconnect the wiring from the fan shroud. Completed ❏

4. Unbolt the fan shroud from the radiator. Completed ❏

5. Pull the fan shroud and fan motor out of the engine compartment. Completed ❏

6. Loosen the hose clamps securing the upper and the lower radiator hoses to the radiator. Remove the upper and lower hoses from the radiator. See **Figure 12-1**. Completed ❏

Figure 12-1. Radiator hose clamps secure the hoses to the radiator or the engine. Two types of hose clamps are shown here. Spring clamp. Screw, band, worm gear clamp.

7. If the vehicle has an automatic transmission, position a drain pan under the transmission coolant lines on the radiator. Disconnect the transmission coolant lines from the radiator. Allow transmission fluid in the lines to drain into the drain pan. Completed ❏

8. Loosen the clamp holding the coolant recovery tank hose to the radiator. Remove the hose from the radiator. Completed ❏

9. Unbolt the radiator. Often, the radiator is secured at the top with bolts and held in place at the bottom by pegs. Once the radiator is unbolted, lift it from the engine compartment. Completed ❏

Radiator Installation

10. As an alternative method of installation, bolt the fan shroud and fan motor to the radiator. Completed ❏

11. Slide the radiator (with the shroud/fan motor assembly attached) into the engine compartment and bolt it into place. Completed ❏

12. Attach the upper and lower radiator hoses to the radiator. Tighten the hose clamps. Completed ❏

13. Install the transmission coolant lines. Completed ❏

14. Install the coolant recovery tank hose to the radiator and tighten the clamp. Completed ❏

15. Plug in the fan motor. Completed ❏

16. Fill the radiator with the recommended type and amount of coolant. Completed ❏

17. Reconnect the negative battery cable. Completed ❏

18. If applicable, check the automatic transmission fluid level and add fluid as needed. Completed ❏

Instructor's Initials _____

Date _____

Name _____ Date _____ Class _____

Job 13

Removing and Installing a Drive Axle

Objective

After completing this job, you will be able to remove and install a drive axle from a front-wheel-drive vehicle.

Equipment and Materials

To complete this job, you will need the following:
- Front-wheel-drive vehicle
- Ratchets, sockets, and wrenches
- CV joint puller or long bar
- Jack stands, floor jack, and wheel chocks or a lift

Safety Notice: Before performing this job, review all pertinent safety information in the text and discuss safety procedures with your instructor.

Procedure

Drive Axle Removal

1. Turn the steering wheel so the front wheels are pointing straight ahead. Completed ❑

2. Safely raise the vehicle with a lift. If a lift is not available, chock the rear wheels and raise the front of the vehicle with a floor jack. Then, lower the front of the vehicle onto jack stands. Completed ❑

3. Remove the front wheels. Completed ❑

4. Locate the drive axle nut. Check the nut to see if there is an indentation (stake) in the nut at the threads. If a stake is present use a punch to straighten the indentation. Completed ❑

5. Loosen the drive axle nut. Completed ❑

6. Separate the lower control arm from the steering knuckle. Completed ❑

7. Loosen the drive axle nut until the nut is at the end of the shaft. Either hit the nut with a hammer to break the outer CV joint housing from steering knuckle or use a CV joint puller. Completed ❑

Copyright by Goodheart-Willcox Co., Inc. May not be reproduced or posted to a publicly accessible website.

8. Remove the drive axle nut. Pull the suspension assembly away from the vehicle while pushing the shaft toward the center of the vehicle to separate the steering knuckle and drive axle. Completed ❑

9. Do not allow the drive axle to hang by the inner CV joint. Use wire to suspend the outer CV joint from the body. What would happen if you let the drive axle hang by the inner CV joint?

_____

_____ Completed ❑

10. Remove the inner CV joint. If a circlip retains the shaft in the transaxle, use a puller or a long bar to pull the joint from the transaxle. If the shaft is bolted to the transaxle flange, remove the bolts. Completed ❑

11. Pull the drive axle out from under the vehicle. Examine the circlip (if used) on the transaxle end of the shaft. Completed ❑

Drive Axle Installation

12. Install the drive axle nut on the end of the shaft. If a circlip is used to retain the axle, make sure the clip is in place and install the inner CV joint in the transaxle by tapping on the nut. You must hear the circlip snap into place. If bolts are used to retain the axle, simply bolt the axle to the transaxle flange. Completed ❑

13. Remove the drive axle nut. Align the splines on the shaft to fit the grooves in the steering knuckle. Slip the drive axle through the steering knuckle. Completed ❑

14. Install the drive axle nut. Completed ❑

15. Reconnect the steering knuckle and lower the control arm. Completed ❑

16. Install the front wheel. Completed ❑

17. If necessary, jack up the vehicle and remove the jack stands. Completed ❑

18. Tighten the drive axle nut and the wheel lug nuts to the specified torque value. Completed ❑

19. Install the hub cap. Completed ❑

Instructor's Initials _____

Date _____

Name _____ Date _____ Class _____

Job 14

Electrical Practice

Objective

After completing this job, you will be able to test for power, voltage, and resistance with a nonpowered test light and a DVOM. You will also be able to splice wires and clean a battery.

Equipment and Materials

To complete this job, you will need the following:
- Vehicle
- Nonpowered test light
- Used door assembly (with power window motor)
- Jumper wires
- Wires to splice together, wire with a socket at one end, or a length of wire
- Soldering iron
- Rosin-core solder
- Heat-shrink tubing
- Heat gun
- Crimping pliers
- Crimp connector
- Baking soda and water
- Wire brush
- Wrenches
- DVOM

Safety Notice: Before performing this job, review all pertinent safety information in the text and discuss safety procedures with your instructor.

Procedure

Testing for Power with a Test Light

1. With the headlight/marker lamp or parking lamp switch off, remove a marker lamp or parking lamp from the vehicle. Completed ❑

2. Remove the bulb and socket from the lamp. Completed ❑

3. Remove the bulb from the socket. Completed ❑

4. Attach the ground wire from a nonpowered test light to a good ground. Completed ❑

5. Insert the nonpowered test light probe into the socket. Does the test light turn on?

   _____ Completed ❑

6. Remove the test light probe from the socket. Completed ❑

7. Turn on the headlight/marker lamp or parking lamp switch. Completed ❑

8. Reinsert the test light probe into the socket. Does the test light turn on?

   _____ Completed ❑

9. What problems could prevent the test light from turning on in step 8?

   _____ Completed ❑

Supplying Power

10. Obtain a used door assembly that contains a power window. Completed ❑

11. Attach a jumper wire long enough to reach the power window motor or wiring harness to the positive clamp of a battery charger. Attach another jumper wire to the negative clamp of the battery charger. Completed ❑

12. Touch one jumper wire to one terminal of the motor and touch the other wire to the other terminal on the power window motor. Position the wires so they will *not* touch each other. Completed ❑

13. Turn on the battery charger. Completed ❑

14. The motor should either try to raise the window or lower it. Reversing the jumper wires will reverse the movement of the window. Completed ❑

Checking a Battery with a DVOM

15. Set the DVOM to read dc volts. Completed ❑

16. With the vehicle engine off, place the red lead on the positive battery post and the black lead on the negative battery post. Completed ❑

17. Read the dc volts on the DVOM. Record the reading here.

    _____ Completed ❑

18. Start the vehicle and again place the meter leads on the battery posts. Record the dc voltage on the DVOM reading here.

    _____ Completed ❑

19. What is the difference in engine-off and engine-on readings?

    _____ Completed ❑

20. What causes the difference in engine-off and engine-on readings?

    _____ Completed ❑

Job 14 Electrical Practice

Name _____

Checking for Continuity with a DVOM

21. Set the DVOM selector switch to the ohms (Ω) setting. Completed ❏

22. Using a length of wire or a wire with a socket at one end, touch one meter lead to one end of the wire. Completed ❏

23. Touch the other meter lead to the other end of the wire, or to the connectors inside the socket. See **Figure 14-1**. Record the reading below.

    _____ Completed ❏

Figure 14-1. A digital multimeter, or DVOM, can be used to check the voltage, current, and resistance in a circuit. Here, it is being used to check a bulb circuit for continuity.

24. What does the reading tell you?

    _____ Completed ❏

Soldering Wires Together

25. Start with two wires of the same diameter. Slip heat-shrink tubing over one of the wires. Completed ❏

26. Strip about 1″ of insulation from the end of each wire. Completed ❏

27. Twist the wires together. Do not allow the wire ends to touch the insulation. Completed ❏

28. Heat the soldering iron and coat the tip with rosin-core solder. Completed ❏

29. Heat the twisted wires with the tip of the soldering iron. Completed ❏

30. Melt solder onto the wire using only the heat from the wire, *not* the heat from the iron. See **Figure 14-2**.   Completed ❏

Figure 14-2. Heat the wires with the tip of the soldering gun and melt solder on the heated wires. Do not heat the solder and drip it onto the wires.

31. Coat all the twisted wires with solder.   Completed ❏
32. Slide the heat-shrink tubing over the bare wire area.   Completed ❏
33. Shrink the heat-shrink tubing with the heat from a heat gun. The heat-shrink tubing will protect the bare wire from contamination.   Completed ❏

Crimping Wires Together

34. Start with two wires of the same diameter.   Completed ❏
35. Strip 1/2″ of insulation from the end of each wire.   Completed ❏
36. Insert the end of one wire into one end of the crimp connector. If all the wire strands will not fit into the crimp connector, twist the wire strands together to allow them to slip inside the connector.   Completed ❏
37. Insert the other wire into the other end of the crimp connector.   Completed ❏
38. Use crimping pliers to crush the crimp connector and join the wires.   Completed ❏
39. If the connector has a plastic shrink-tubing cover, heat the cover to create an airtight connection.   Completed ❏

Cleaning a Battery

40. Disconnect the negative battery cable from the battery. Position the cable end so it cannot accidently touch the negative terminal.   Completed ❏
41. Disconnect the positive battery cable and lay it to the side.   Completed ❏
42. Locate and unbolt the battery hold-down.   Completed ❏
43. Remove the battery from the vehicle.   Completed ❏

Job 14 Electrical Practice

Name _____

44. Inspect the battery case. Look for cracks, dirt, and corrosion. Dirt and corrosion can be cleaned from the case. A cracked case requires battery replacement. Completed ❑

45. Clean the case with a mixture of baking soda and water. Then rinse the case with clean water. Completed ❑

46. Look for corrosion in the battery tray. If corrosion is present, clean the tray with baking soda and water and rinse it with clean water. Completed ❑

47. Look for corrosion on the battery cable ends and the battery terminals. Completed ❑

48. Remove corrosion from the terminals and cable ends with a wire brush. If there is corrosion within the cables, replace the cables. Completed ❑

49. Install the battery on the battery tray and connect the positive cable end to the positive battery terminal. Make sure it is tight. Completed ❑

50. Connect the negative cable end to the negative battery terminal. Make sure it is tight. Completed ❑

Instructor's Initials _____

Date _____

Name _____ Date _____ Class _____

Job 15

Masking

Objective

After completing this job, you will be able to apply masking tape and masking paper, back tape, raise a molding with rope, and mask a door opening and windshield.

Equipment and Materials

To complete this job, you will need the following:
- 3/4", 1/4", and 2" masking tape
- 6" and 18" masking paper
- Plastic bone
- Masking rope
- Pre-paint solvent
- Vehicle with tape stripes

Safety Notice: Before performing this job, review all pertinent safety information in the text and discuss safety procedures with your instructor.

Procedure

Practice Tape Application

1. Start with a clean surface. Wash a fender with soap and water, and dry the panel thoroughly. Why do so many collision repair tasks start by cleaning the panel?

   _____

   _____

   Completed ❑

2. Pull a 1" section of 3/4" masking tape loose from the roll and attach it to the fender on the edge of the stripe or body line. Hold the tape in place and pull the roll. Apply at least 2' of masking tape along the stripe or body line.

   Completed ❑

3. Apply 1/4" masking tape next to the 3/4" tape. Work carefully so there is no gap between the 1/4" and 3/4" tapes.

   Completed ❑

Copyright by Goodheart-Willcox Co., Inc. May not be reproduced or posted to a publicly accessible website.

4. Apply 2″ masking tape next to the 1/4″ tape. Leave no gap between the two pieces of tape. Did you have difficulty applying the 1/4″ or 2″ masking tape? If so, explain.

_____

_____   Completed ❏

Practice Back Taping

5. Open the hood. Clean the fender flange and the inside edge or flange of the hood with soap and water. Dry the panels thoroughly.   Completed ❏

6. Pull a 1″ section of 3/4″ masking tape loose from the roll and attach it to the edge of the fender so the middle of the tape is right on the edge of the flange.   Completed ❏

7. Hold the tape in place and pull back on the roll to dispense more tape and apply it along the edge of the fender in the same manner. Apply at least 2′ of masking tape.   Completed ❏

8. Apply masking tape to the inside edge of the hood in the same way.   Completed ❏

9. Roll the tape on the fender and hood to make a soft edge.   Completed ❏

10. Open the door. Clean the inside of the door and the door opening with soap and water. Dry these areas thoroughly.   Completed ❏

11. Apply 6″ masking paper and 3/4″ masking tape to the entire rear edge of the inside of a door. Leave a small amount of the tape extending beyond the edge of the door. Roll the protruding edge back.   Completed ❏

12. Apply 6″ masking paper and 3/4″ masking tape to the entire door opening edge, including the rocker panel. As before, position the tape so a small amount of it extends beyond the edge. Roll the protruding edge back.   Completed ❏

Raising Windshield Moldings

13. Use a bone to separate the windshield molding from the roof. Insert the bone between the molding and the paint, and move the bone all the way around the molding.   Completed ❏

14. Use soap and water to clean the area under the molding.   Completed ❏

15. Wipe the area with pre-paint solvent and dry it with a clean paper towel.   Completed ❏

16. Use the bone to push masking rope under the molding. The masking rope will raise the molding and allow the roof to be painted without creating a visible paint edge at the molding.   Completed ❏

Practice Masking a Windshield

17. Outline the windshield molding with 3/4″ tape. The purpose of this task is to develop accuracy in taping, so be careful to stay just along the edge.   Completed ❏

18. Tear off a sheet of 18″ masking paper that is long enough to span the width of the windshield.   Completed ❏

19. To cover the upper portion of the windshield, apply a sheet of 18″ masking paper to the tape already applied around the upper molding.   Completed ❏

Job 15 Masking 287

Name _____

20. Fold the corners of the masking paper as needed to conform to the sides of the windshield. Completed ❑

21. Tape the side edges to the masking tape already applied to the molding. Completed ❑

22. To cover the lower portion of the windshield, tape another sheet of 18″ masking paper to the first sheet of 18″ masking paper. Completed ❑

23. Fold and tape the edges of the second sheet as required. See **Figure 15-1**. Completed ❑

Figure 15-1. Two pieces of 18″ masking paper were used to cover this windshield opening.

24. Apply tape to the masking paper to cover and seal any loose folds or edges. Why is this final step important?

_____
_____ Completed ❑

Instructor's Initials _____

Date _____

Copyright by Goodheart-Willcox Co., Inc. May not be reproduced or posted to a publicly accessible website.

Name _____ Date _____ Class _____

Job 16

Mixing, Applying, and Sanding Body Filler

Objective

After completing this job, you will be able to mix, apply, cure, and block sand body filler on a variety of panel contours.

Equipment and Materials

To complete this job, you will need the following:
- Practice panels containing low-, high-, and double-crown areas
- Panel with a 12″ × 12″ bare metal area
- Body filler mixing board
- Body filler
- Hardener
- Body filler spreader(s)
- Cheese grater
- Sanding blocks
- 40-, 80-, and 180-grit sandpaper

Safety Notice: Before performing this job, review all pertinent safety information in the text and discuss safety procedures with your instructor.

Procedure

Mixing Body Filler

1. Read the mixing instructions on the can of body filler. Summarize the instructions below.

   _____
   _____
   _____
   _____

   Completed ❑

2. Remove a golf ball size amount of body filler from the can and place it on the mixing board. Why should a mixing board be used instead of a piece of cardboard?

_____

_____ Completed ❑

3. How much hardener should you add to this amount of body filler?

_____ Completed ❑

4. Squeeze out the proper amount of hardener. See **Figure 16-1**. Stir the hardener into the body filler with the mixing stick. Stir the filler until it is a uniform color (no streaks), but do not overdo it. Vigorously stirring the filler will create air bubbles in the hardened filler. Work quickly because the filler will begin to cure as soon as the hardener is added. Completed ❑

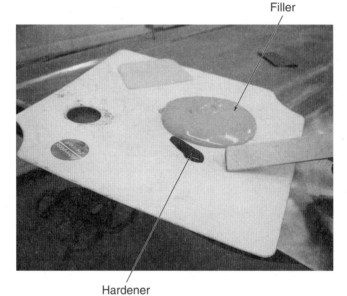

Figure 16-1. Always use clean mixing sticks, paddles, and mixing boards. Dirty equipment can contaminate the newly mixed filler. The mixing board must have a nonabsorbent surface. Never use cardboard as a mixing board.

Applying Body Filler

5. Use a spreader to apply the mixed body filler to the bare metal panel. Press down firmly on the spreader to set the body filler in place on the panel Completed ❑

6. Spread the body filler until you have made a uniform 1/4″ thick patch. Completed ❑

7. Use the spreader to make the surface of the body filler as smooth as possible. Completed ❑

8. Stop spreading when the body filler begins to harden. Completed ❑

9. Feel the back side of the panel in the body filler application area. What do you notice?

_____

_____

_____ Completed ❑

Job 16 Mixing, Applying, and Sanding Body Filler

Name _____

Cheese Grating Plastic Body Filler

Body filler must cure before it can be sanded. Under-cured body filler will not adhere properly to the panel. This lack of adhesion is most pronounced along the featheredge. Semi-cured body filler can be cheese grated if you are careful not to shred the edge. Cheese grating is an optional step, as sanding will also remove filler. However, cheese grating is faster than sanding and can be done earlier in the cure cycle. Be aware that once the body filler has cured completely, it is too late to cheese grate.

10. Once the body filler has cured to the consistency of processed cheese, it can be cut down with a cheese grater. If the panel has a crown, grate perpendicular to the crown. Avoid the edges of the body filler. See **Figure 16-2**. Completed ❑

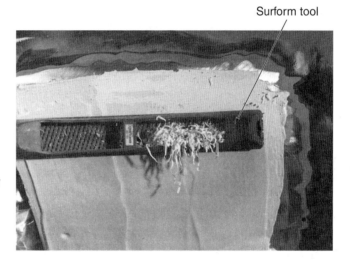

Figure 16-2. The surform tool, also known as a cheese grater, works like a file to remove excess filler. File when the filler has hardened to the consistency of processed cheese. Do not wait too long. Filler is difficult to shape once it has hardened completely.

11. Move the cheese grater in one direction only. Shave off a few strands of body filler with each stroke. The goal is to remove any high ridges in the body filler with the cheese grater. Do not remove more than 1/8″ of the body filler during this procedure. Completed ❑

Sanding Plastic Body Filler

12. Allow the filler to cure completely. Select a sanding block that will span the width of the body filler. Completed ❑

13. Block sand with 40-grit sandpaper to smooth the body filler to a uniform 1/16″ thickness. If the panel has a crown, move the sanding board up and down *with* the crown. Completed ❑

Contour Block Sanding

14. Use the rounded end of a ball peen hammer to tap a small dent (less than 1/8″ deep) in the low-crown, high-crown, and double-crown areas on the panel. Completed ❑

15. Remove the paint from each dent. There should be at least 2″ of bare metal on all sides of the dents. Completed ❑

16. Mix and apply an overfill of body filler in the dents. Overfill means that more body filler is applied than is required to fill the dent. Completed ❑

17. While you wait for the filler to cure, answer the following questions.

    What type of sanding block will you use to shape the filler and how will you hold the block to sand the following contours?

    Low-crown area:

    _____
    _____

    High-crown area:

    _____
    _____

    Double-crown area:

    _____
    _____ Completed ❑

18. Block sand the body filler with 80-grit sandpaper. Use the contour of the undamaged portion of the panel as a guide for the sanding block. Once you have sanded a featheredge around the entire patch of body filler, switch to 180-grit sandpaper on the sanding block. Complete the block sanding by sanding out the deep 80-grit scratches in the body filler with 180-grit sandpaper. Completed ❑

19. If you over sand and undercut the body filler, mix and apply more body filler. Block sand the patch until the filler matches the original (undamaged) contour of the panel. This may take several attempts. Keep trying until you get it right. Remember, the goal is to fill and block sand the dent in one attempt. This is an acquired skill. It takes practice to reach this goal. Completed ❑

20. Which contour was the most difficult panel to block sand? Explain why.

    _____
    _____
    _____
    _____ Completed ❑

Instructor's Initials _____

Date _____

Name _____ Date _____ Class _____

Job 17

Reducing Paint

Objective

After completing this job, you will be able to reduce paint using a mixing stick or mixing cup.

Equipment and Materials

To complete this job, you will need the following:
- Water tinted with blue food coloring to represent basecoat. Water tinted with red food coloring to represent hardener. Clear water to represent reducer.
- Mixing sticks (standard and dedicated)
- Paint stick
- Tape measure
- Marker or pen
- Straight-sided container
- Mixing cup

Safety Notice: Before performing this job, review all pertinent safety information in the text and discuss safety procedures with your instructor.

Procedure

Using a Mixing Stick

1. Place a standard mixing stick into the straight-sided container. Why will a mixing stick work with a straight-sided container only?

   _____

   _____ Completed ❏

2. Add basecoat (blue water) to the container until it reaches the "1" mark on the mixing stick. Completed ❏

3. Add the same amount of reducer (red water) to the container. Completed ❏

4. What is this mixing ratio?

   _____ Completed ❏

5. Empty the container. Completed ❏

6. Add two parts basecoat and one part reducer to the container. Completed ❏

7. What is this mixing ratio?

   _____   Completed ❏

8. Empty the container.   Completed ❏

9. Study the dedicated mixing stick. What is the ratio?

   _____   Completed ❏

10. Place the dedicated mixing stick in a straight-sided container.   Completed ❏

11. Add the basecoat to the level of the first 3 on the mixing stick, reducer to the next 3, and hardener to the last 3.   Completed ❏

12. Empty the container.   Completed ❏

13. Make your own 4:2:1 mixing stick by marking increments on a paint stick as shown in **Figure 17-1**. Make each part 1/2″ in length.   Completed ❏

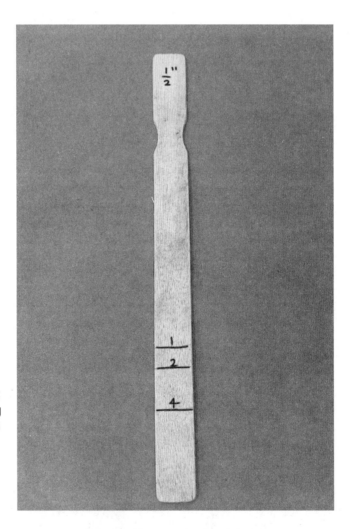

Figure 17-1. This paint stick has been made into a mixing stick by marking graduations on it. It can be used to produce a 4:2:1 ratio. Use a ruler to mark 1/2″ units on a paint stick. Each one of the 1/2″ units is a "part."

14. Place the stick in the container and add basecoat to the 4, reducer to the 2, and hardener to the 1.   Completed ❏

Job 17 Reducing Paint

Name _____

Using a Mixing Cup

Mixing cups usually have several ratios to choose from.

15. Select the ratio of 1:1. Add basecoat and reducer at this ratio to the *highest level* available. Empty the cup. .... Completed ❑

16. Select the ratio of 4:2:1. Add basecoat, reducer, and hardener at this ratio to the *lowest level* available. Empty the cup. .... Completed ❑

17. Select the ratio of 8:4:1. Add basecoat, reducer, and hardener at this ratio to the *highest level* available. Empty the cup. .... Completed ❑

Instructor's Initials _____

Date _____

Name _____ Date _____ Class _____

Job 18

Spray Gun Basics

Objective

After completing this job, you will be able to demonstrate proper spray gun handling techniques, including body position, distance, speed, fan orientation, overlap, and triggering.

Equipment and Materials

To complete this job, you will need the following:
- Spray gun
- 8 1/2″ × 11″ piece of cardboard
- Hose clamp
- Ruler or tape measure
- Scissors or razor knife
- Clock or watch
- Masking tape
- Vehicle hood and fender mounted on saw horses, or a vehicle
- Appropriate safety gear

Safety Notice: Before performing this job, review all pertinent safety information in the text and discuss safety procedures with your instructor.

Procedure

1. Practice maintaining the correct body position. Position your body on the left side of the hood if you are right-handed or on the right side of the hood if you are left-handed. Stand near the front of the hood, not in the center. Place your feet about 3′ apart. Completed ❑

2. Good painters are flexible. Stretch to reach the middle and rear of the hood. See **Figure 18-1**. Practice stretching, but do not lean so much that your body touches the edge of the hood. Where should the air hose be held while a painter is stretching?

   _____

   _____ Completed ❑

Figure 18-1. A paint technician must be able to stretch and squat when spraying. Flexibility, eye/hand coordination, and rhythm are all part of good spray technique.

3. Cut a triangular spray gun fan pattern from a piece of cardboard. Leave a square at the top of the pattern so the triangle can be attached to the spray gun air cap with a hose clamp. See **Figure 18-2**. The triangle should be 10″ at the base and 8″ tall. Completed ❑

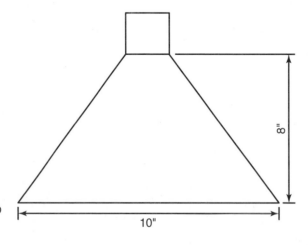

Figure 18-2. Use a cardboard cutout with the dimensions shown to represent the spray fan.

Job 18 Spray Gun Basics

Name _____

4. Attach the cardboard triangle to the spray gun air cap with a hose clamp. Hold the spray gun so the cardboard triangle touches the hood. See **Figure 18-3**. The 8″ gap between the panel and the spray gun air cap is the distance you want to maintain while painting. Try to maintain the 8″ gap while practicing on several different areas of the hood. What is arcing?

_____

_____   Completed ❑

Figure 18-3. For consistent paint coverage, the fan must strike the surface uniformly. Note how the technician's wrist is bent to properly orient the fan.

5. Practice maintaining proper fan orientation. Keep the cardboard triangle flat on all areas of the hood. Notice how you must bend your wrist to keep the base of the cardboard triangle flat on the hood. You must constantly reposition the spray gun while you apply paint. What is heeling?

_____

_____   Completed ❑

6. Develop the skill of maintaining a constant spray gun travel speed while practicing your strokes. You want to move at a steady rate of one foot-per-second. For example, if the hood is 4′ long, you want each stroke to last four seconds. Stretch to reach the rear of the hood. Practice moving your legs at the knees instead of just moving your arm. Your legs have much more strength and endurance than your arms. Once your arm tires, it becomes difficult to maintain a consistent distance and speed. Practice until your spray gun travel speed is steady.   Completed ❑

7. Once you can make a stroke with the proper distance, speed, and fan orientation, practice maintaining proper overlap. Use masking tape to make lines down the length of the hood. The first line should be at the edge of the hood, and the other lines should be 5″ apart. The masking tape represents the center of a stroke. Position the center of the cardboard triangle base on the masking tape. Practice following the tape lines at the proper speed with the base of the cardboard fan always evenly touching the hood.   Completed ❑

8. Once you have mastered distance, speed, fan orientation, and overlap, practice proper triggering technique. Continue making practice strokes, but this time, begin the stroke off the hood. As you begin your stroke, pull the trigger all the way back before the cardboard fan reaches the hood. Continue the stroke until the spray gun has moved past the hood before you release the trigger. Practice pulling the trigger at the beginning of the stroke and releasing the trigger at the end of the stroke. Completed ❑

9. Once you have had sufficient practice on the hood, move to the fender. Position your body at the front of the fender and stretch to reach the rear of the fender. As with the hood, hold the cardboard triangle flat against the fender. Notice how often you must turn the spray gun to keep the base of the triangle flat against the panel. You will need to kneel or squat to reach the lower part of the fender. Practice body position, distance, speed, fan orientation, overlap, and triggering on the fender. Completed ❑

Instructor's Initials _____

Date _____

Name _____ Date _____ Class _____

Job 19

Spray Gun Practice

Objective

After completing this job, you will be able to demonstrate proper spray gun handling techniques while spraying water.

Equipment and Materials

To complete this job, you will need the following:
- Spray gun
- Ruler or tape measure
- Clock or stopwatch
- Compressed air hose and supply of compressed air
- Vehicle hood and fender mounted on saw horses (or on a vehicle)
- Appropriate safety gear

Safety Notice: Before performing this job, review all pertinent safety information in the text and discuss safety procedures with your instructor.

Procedure

1. Fill the spray gun cup with water. Hook up the compressed air hose to the air inlet fitting on the spray gun. Make the following basic adjustments to the spray gun: set the fan wide open, turn the material adjustment knob to expose one thread, and set the air pressure to 10 psi at the inlet for HVLP spray guns and 35 psi at the inlet for conventional spray guns. Completed ❑

2. Practice proper body position. Position yourself at the front of the hood and stretch to make sure you can reach the rear center of the hood. Hold the air hose behind your back with one hand. The air hose must *never* touch the hood or any panel you paint. Keeping the air hose out of the way takes considerable concentration. Completed ❑

3. Make a few practice strokes to get used to spraying water. When you are ready, measure 6″–8″ up from the hood with a ruler or tape measure. Practice maintaining proper gun distance by positioning the air cap of your spray gun at this 6″–8″ distance. Always try to keep the spray gun at this distance from the panel. Completed ❑

4. Practice maintaining proper gun speed. The correct travel speed is 1 foot per second. Practice moving the spray gun at this speed while maintaining the proper gun distance. Keep your arm straight. Use your legs to shift the position of your body and the spray gun. The arms are considerably weaker than the legs. You will find that it is easier to maintain consistent speed if you incorporate using your legs as you stretch your body from one end of the panel to the other. See **Figure 19-1.** Completed ❑

Figure 19-1. Proper positioning and stretching will help the painter maintain a consistent distance and speed.

5. Practice maintaining the proper fan orientation. Remember how you had to bend your wrist in Job 18? The spray pattern must hit the panel evenly to apply a smooth coat of paint. You must avoid heeling and arcing! You must bend your wrist for uniform paint application. As you make strokes at a constant distance and speed, practice keeping the spray pattern even. Completed ❑

6. Once you can make a stroke of uniform distance and speed with the proper fan orientation, practice maintaining a 50% overlap. Aim the center of the first stroke at the edge of the hood, then aim the center of all subsequent strokes at the edge of the previous stroke. Work your way across the hood. Completed ❑

7. Once you have mastered distance, speed, fan orientation, and overlap, practice triggering. Pull the trigger all the way back at the beginning of the stroke, and release the trigger at the end of the stroke. Completed ❑

8. After you have mastered spray gun technique on the hood, move to the fender. Position yourself at the front of the fender. Make sure you can stretch to reach the entire fender. Your first stroke should follow the arch of the wheel opening. Then make strokes starting at the top of the fender and working down. When you reach the arch, spray the front of the fender from the top of the arch to the bottom. Then move to the rear of the fender and work your way down. Completed ❑

Instructor's Initials _____

Date _____

Name _____ Date _____ Class _____

Job 20

Spraying Basecoat/Clearcoat Paints

Objective

After completing this job, you will be able to apply basecoat and clearcoat to vertical and horizontal panels.

Equipment and Materials

To complete this job, you will need the following:
- Clean spray booth
- Spray gun
- Tack rag
- Basecoat and reducer
- Clearcoat and hardener
- Prepped fender and hood

Safety Notice: Before performing this job, review all pertinent safety information in the text and discuss safety procedures with your instructor.

Procedure

Preparing the Air System

1. Check the air line, air regulator, air hose, fittings, and quick-connects for leaking air. Completed ❏

2. Clean dust from the air hose by wiping it with a tack rag. Why should you clean dust from the air hose?

   _____

   _____ Completed ❏

Spray Gun Preparation

3. What type of spray gun will be used for this job?

   _____ Completed ❏

4. Remove the spray gun's air cap, material adjustment knob, spring, needle, and fluid tip. Check the internal paint passages in the spray gun for dried paint. Clean as needed. Reassemble the spray gun. Completed ❏

303

5. List the size of the fluid tip.

   _____ Completed ❏

6. What size pattern, in inches, will the spray gun be set at?

   _____ Completed ❏

7. How many turns out on the material adjustment knob are required to achieve the proper pattern for applying basecoat?

   _____ Completed ❏

8. How many turns out on the material adjustment knob are required to achieve the proper pattern for applying clearcoat?

   _____ Completed ❏

9. What will the air pressure, measured at the spray gun inlet, be set at for applying basecoat?

   _____ Completed ❏

10. What will the air pressure, measured at the spray gun inlet, be set at for applying clearcoat?

    _____ Completed ❏

11. List safety concerns and considerations when spraying basecoat/clearcoat.

    _____
    _____
    _____
    _____ Completed ❏

Basecoat Preparation

12. Is the basecoat you will be applying solvent-borne or waterborne?

    _____ Completed ❏

13. What is the mixing ratio for the basecoat?

    _____ Completed ❏

14. How long should the basecoat be stirred?

    _____ Completed ❏

15. Some paints cure by polymerization, which is affected by spray booth temperature. What is the temperature of the spray booth?

    _____ Completed ❏

16. What type of reducer will be added to the basecoat?

    _____ Completed ❏

17. Add the correct amount of reducer to the basecoat. Completed ❏

18. What is the recommended viscosity of the basecoat?

    _____ Completed ❏

Job 20 Spraying Basecoat/Clearcoat Paints

Name _____

19. Measure the viscosity. How does the measured viscosity compare to the recommended viscosity?

   _____ Completed ❏

20. Strain the basecoat as it is added to the spray gun. Completed ❏

Basecoat Spray Technique

21. How far will the spray gun be held from the surface?

   _____ Completed ❏

22. How fast will the spray gun be moved?

   _____ Completed ❏

23. How much will each stroke overlap the previous stroke?

   _____ Completed ❏

Applying Basecoat

24. How many coats of basecoat are required?

   _____ Completed ❏

25. What is the flash time between coats?

   _____ Completed ❏

26. Is a drop coat required?

   _____ Completed ❏

27. Hook up the spray gun to the air line in the booth. See **Figure 20-1**. Completed ❑

Figure 20-1. With air flowing through the spray gun, set the air pressure at the regulator on the spray booth wall.

28. Adjust the spray gun. Completed ❑
29. Unfold a tack rag and wad it up into a loose ball. Tack off the hood and fender. Completed ❑
30. Apply the first coat of basecoat. Completed ❑
31. Evaluate the first coat. If you find a run, quickly wipe off the run with a rag soaked in wax-and-grease remover. If fish eyes are present, hold the spray gun about 12″ to 15″ away from the panel and spray a dry coat of basecoat over the fish eyes. If dirt or contamination is present, allow the basecoat to dry and sand out the contamination with 1000-grit sandpaper, using wax-and-grease remover as a lubricant. Completed ❑
32. After the flash time has passed, tack off the hood and fender. Completed ❑
33. Apply the second coat of basecoat. Completed ❑
34. Evaluate the second coat and list any problems found.

_____
_____
_____
_____ Completed ❑

35. After the flash time has passed, tack off the hood and fender. Completed ❑
36. Apply the third coat of basecoat. Completed ❑

Job 20 Spraying Basecoat/Clearcoat Paints

Name _____

37. Evaluate the third coat and list any problems found. Look closely to make sure the undercoat is completely covered. If mottling is present, apply a drop coat. A drop coat is sprayed at one-half normal air pressure.   Completed ❑

38. Clean the spray gun.   Completed ❑

Applying Clearcoat

39. What is the mixing ratio for clearcoat?
_____   Completed ❑

40. What type of hardener is added to the clearcoat?
_____   Completed ❑

41. How long should you stir the clearcoat?
_____   Completed ❑

42. Add the proper amount of hardener to the clearcoat.   Completed ❑

43. What is the recommended viscosity of the clearcoat?
_____   Completed ❑

44. Measure the viscosity of the clearcoat. Explain how to change viscosity.
_____
_____
_____   Completed ❑

45. Strain the clearcoat as it is added to the spray gun.   Completed ❑

46. How many coats of clearcoat are required?
_____   Completed ❑

47. What is the flash time between clearcoats?
_____   Completed ❑

Clearcoat Spray Technique

48. How far will the spray gun be held from the surface when spraying clearcoat?
_____   Completed ❑

49. How fast will the spray gun be moved when spraying clearcoat?
_____   Completed ❑

50. How much will each stroke overlap the previous stroke?
_____   Completed ❑

Clearcoat Application

51. How long is the flash time between basecoat and clearcoat application?

    _____ Completed ❑

52. Hook up the spray gun to the air line in the booth. Completed ❑

53. Adjust the spray gun. Completed ❑

54. Tack off the basecoat and apply the first coat or clear to the hood and fender. Completed ❑

55. Evaluate the clearcoated panel for orange peel or dry spray. If orange peel is present, increase the air pressure by 5 psi for the next coat. If dry spray is present, correct it on the next coat by thoroughly wetting every part of the panel with clearcoat. Dry spray usually means a missed overlap. Describe any problems encountered.

    _____
    _____
    _____
    _____ Completed ❑

56. After the flash time has elapsed, spray the second coat of clearcoat. This second coat is often difficult to spray because you cannot readily see the edge of the spray when overlapping. Completed ❑

57. If a third coat of clearcoat is required, observe the recommended flash time before spraying. Completed ❑

58. Inspect the final coat of clearcoat. Note any problems here.

    _____
    _____
    _____ Completed ❑

59. Clean the spray gun. Completed ❑

60. In the space below, list any problems you encountered.

    _____
    _____
    _____
    _____
    _____ Completed ❑

Instructor's Initials _____

Date _____

Name _____ Date _____ Class _____

Job 21

Preparing for a Clearcoat Blend

Objective

After completing this job, you will be able to clean a panel prior to sanding and use three methods to prepare a panel for a clearcoat blend.

Equipment and Materials

To complete this job, you will need the following:
- Demonstration hood with an OEM clearcoat
- Car wash soap and hot water
- Wax-and-grease remover
- Clean paper towels
- DA sander with soft backing pad
- Gray scuff pad
- Sanding paste
- 1000-grit sandpaper
- Sanding pad

Safety Notice: Before performing this job, review all pertinent safety information in the text and discuss safety procedures with your instructor.

Procedure

You will prepare one section of the hood with a DA sander and 800-grit sandpaper, a second section of the hood with a gray scuff pad and sanding paste, and a third section of the hood with 1000-grit wet sandpaper. An OEM clearcoat surface is not flat. As you sand, you will notice shiny dots remaining on the surface. These are low areas in the clearcoat's orange peel. Continue sanding to eliminate these shiny dots and level the surface. The refinish clearcoat will look much better if it is applied over a level surface.

1. Wash the hood with hot, soapy water and dry it off. Is it acceptable to use laundry detergent, or should you use soap specifically made to wash automobiles?

_____

_____

Completed ❑

2. Clean the panel with wax-and-grease remover. If you use a spray bottle, wet a 12" × 12" area. While the surface is wet, wipe off the wax-and-grease remover with clean paper towels. Do not allow the wax-and-grease remover to dry. If you do not have a spray bottle, soak a clean paper towel and wet a 12" × 12" area. Wipe off the wet wax-and-grease remover with clean paper towels. Continue wetting and wiping until the entire panel is clean.

Completed ❑

3. Use masking tape to divide the hood into three sections as shown in **Figure 21-1**.

Completed ❑

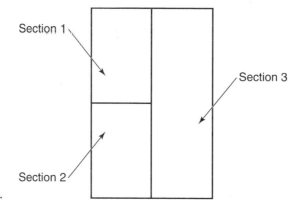

Figure 21-1. Use masking tape to divide the hood into three sections as shown here.

4. Use 800-grit dry sandpaper on a DA sander equipped with a soft pad to scuff the clearcoat on Section 1 of the hood. Keep the DA sander flat and do not push down. See **Figure 21-2**. Avoid sanding body lines and panel edges. Scuff one 6" × 6" area at a time.

Completed ❑

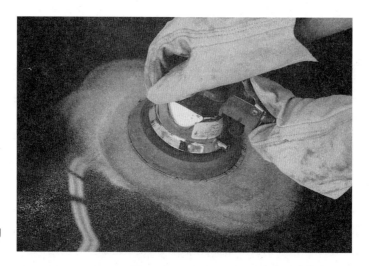

Figure 21-2. Hold the DA sander flat to prevent gouging the surface.

5. Evaluate the scuffing. There should be no shiny areas. The clearcoat should be uniformly dull. If shiny areas are present, continue scuffing until no shiny dots remain on the surface.

Completed ❑

Job 21 Preparing for a Clearcoat Blend

Name _____

6. Use a gray scuff pad and sanding paste to scuff Section 2 of the hood. Wet the scuff pad with water first. Place a 3″ long ribbon of scuff paste on the panel and work the surface in a circular motion with the scuff pad. Scuff one 6″ × 6″ area at a time. You are looking for a uniform dullness. If shiny areas are present, continue scuffing until there are no shiny dots remaining on the surface. Completed ❑

7. Scuff Section 3 of the hood using 1000-grit sandpaper and water. Soak the sandpaper in water before scuffing. Fold the sandpaper into thirds and wrap the sandpaper around a sanding pad. Scuff with a circular motion. Work one 6″ × 6″ area at a time. Remove sanding sludge with a squeegee. Your goal is to scuff until you achieve a uniform dullness. If shiny areas are present, continue scuffing. There should be no shiny dots remaining on the panel surface. Completed ❑

Instructor's Initials _____

Date _____

Name _____ Date _____ Class _____

Job 22

Buffing

Objective

After completing this job, you will be able to apply compound and operate a buffer.

Equipment and Materials

To complete this job, you will need the following:
- Electric buffer with adjustable speed settings
- Color-sanded demonstration panels
- Foam and wool pads
- Buffing compound
- Buffer pad cleaning tool

Safety Notice: Before performing this job, review all pertinent safety information in the text and discuss safety procedures with your instructor.

Procedure

1. Review the following buffing rules before using the buffer to complete this job:
   - Know where the electric cord is at all times. Keep the cord over your shoulder when buffing a horizontal surface like a hood. When buffing a vertical surface, such as a door, keep the cord behind you.
   - Hold the buffer at a low angle, almost flat to the panel. Tilting the buffer up at a steep angle puts too much pressure on a small area. This may concentrate too much heat and burn the paint, or it may cause too much cut and result in a rub through.
   - Avoid contact with panel edges, body lines, antennas, rear view mirrors, or any other surface the spinning buffing wheel can grab.
   - Always keep compound between the buffing pad and the paint.
   - Buff a small area (about 2' × 2') at a time.
   - Keep the buffer moving with a back-and-forth motion. See **Figure 22-1**.
   - Pushing down on the buffer will increase the amount of cut.

   Completed ❑

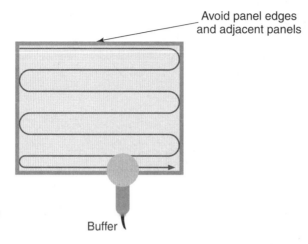

Figure 22-1. Use long back-and-forth strokes to buff a flat panel. The buffer must contact all the paint surfaces except for the edges.

Buffing with a Wool Pad

2. List the type of wool pad (coarse or fine) and the type of compound that will be used for this portion of the job.

   Type of pad: _____

   Type of compound: _____ Completed ❏

3. Apply the compound to a 1′ × 1′ area. Completed ❏

4. Set the buffer at 1200 rpm. See **Figure 22-2**. Completed ❏

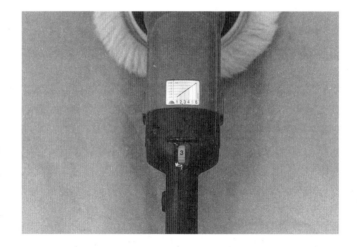

Figure 22-2. Pad rotation speed is adjusted on this buffer by turning the thumb wheel. In this example, the wheel is set to 3. A chart above the wheel gives the pad rotation speed for each number.

5. Use light pressure to buff the 1′ × 1′ area. The pressure you exert on the buffer should not change its rotational speed (1200 rpm). Completed ❏

6. Buff until shine is restored to the paint. Completed ❏

7. Apply compound to another 1′ × 1′ area. Completed ❏

8. Push down on the buffer. Apply medium pressure to the pad and buff the area. The buffer's rotation speed should slow down slightly from the medium pressure. Buff until shine is restored to the paint. Completed ❏

Job 22 Buffing 315

Name _____

9. Compare the cut produced when exerting light pressure to the cut produced when exerting medium pressure.

   Which method cuts faster?
   _____

   Which method is more likely to cause a problem?
   _____   Completed ❑

10. How do you know when a pad should be cleaned?
    _____
    _____   Completed ❑

11. Explain how to clean a wool pad.
    _____
    _____
    _____   Completed ❑

Buffing with a Foam Pad

12. List the type of foam pad (fast cut, medium cut, or low cut) and the type of compound that will be used for this portion of the job.

    Type of pad: _____

    Type of compound: _____   Completed ❑

13. Apply compound to a 1′ × 1′ area.   Completed ❑

14. Set the buffer speed at 1000 rpm.   Completed ❑

15. Use light pressure to buff the area. The buffer rotation speed should not change from the pressure.   Completed ❑

16. Buff until the shine is restored in the paint.   Completed ❑

17. Apply compound to another 1′ × 1′ area.   Completed ❑

18. Use medium pressure to buff the area. The buffer rotation speed should slow down slightly from the pressure. Buff until the shine is restored in the paint.   Completed ❑

19. Compare the cut produced when using light pressure to that produced when using medium pressure.

    Which one cuts faster?
    _____

    Which one is more likely to cause a problem?
    _____   Completed ❑

20. How do you know when it is time to clean the foam pad?

   _____
   _____   Completed ❏

21. Explain how to clean a foam pad.

   _____
   _____
   _____   Completed ❏

Wool vs. Foam Pads

22. Which type of pad do you prefer and why?

   _____
   _____
   _____   Completed ❏

23. Which pad cuts faster?

   _____   Completed ❏

24. Which pad produces a better finish?

   _____   Completed ❏

Instructor's Initials _____

Date _____

Name _____ Date _____ Class _____

Job 23

Applying Pinstripes

Objective

After completing this job, you will be able to apply tape pinstripes and paint pinstripes.

Equipment and Materials

To complete this job, you will need the following:
- Demonstration vehicle
- Soap and water
- Wax-and-grease remover
- Stripe tape
- Razor blade
- 1/8″ and 1/4″ fine-line masking tape
- 3/4″ masking tape
- Fine bristle brush
- Stripe paint

Safety Notice: Before performing this job, review all pertinent safety information in the text and discuss safety procedures with your instructor.

Procedure

Pinstripe Tape Practice

1. Clean the area to be pinstriped with soap and water, and then with wax-and-grease remover. Completed ❑
2. Cut off a 1′ section of pinstripe tape. Completed ❑
3. Remove the paper backing from the tape. Completed ❑
4. Lightly apply the pinstripe tape to the cleaned surface. Completed ❑

317

5. Press the pinstripe tape in place. See **Figure 23-1**. Completed ❏

Figure 23-1. After positioning the pinstripe, it should be pressed into place as shown here.

6. Remove the clear film covering from the pinstripe tape. Completed ❏
7. Again, press the pinstripe tape in place. Completed ❏
8. Remove the practice pinstripe tape. Completed ❏

Applying a Straight Stripe

9. Clean the area where the pinstripe tape will be installed with soap and water. Then wipe it down with wax-and-grease remover. Completed ❏
10. Measure and cut enough pinstripe tape to span one side of the vehicle. Completed ❏
11. Remove a 1′ section of the paper backing and, starting at one end of the vehicle, apply the pinstripe tape at the proper height. Completed ❏
12. Remove the paper backing from the entire piece of pinstripe tape and move to the opposite end of the vehicle. Completed ❏
13. Pull the pinstripe tape tight to make a straight line, but do not pull on the tape so hard that it stretches. Lightly apply the pinstripe tape to the vehicle. Completed ❏
14. Step away from the vehicle and sight down the pinstripe to see if it is straight. Completed ❏
15. Reposition the pinstripe tape if it is not level or straight. Completed ❏
16. When the pinstripe tape is positioned correctly, press down on the tape to lock it in position. Completed ❏
17. Remove the clear film covering from the pinstripe tape. Completed ❏
18. Press the pinstripe tape into position again. Completed ❏

Job 23 Applying Pinstripes 319

Name _____

19. Use a razor blade to cut the tape at the door openings. See **Figure 23-2**. Completed ❑

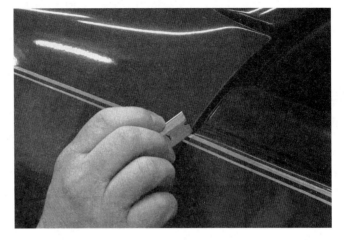

Figure 23-2. Use a razor blade to cut the pinstripe tape at panel openings. After cutting the tape, fold the edges back to the underside of the panel.

20. Fold the pinstripe tape around the door edges. Completed ❑

Applying a Contoured Stripe

21. Clean the area around a wheel opening with soap and water. Then wipe it down with wax-and-grease remover. Completed ❑
22. Measure and cut enough pinstripe tape to go around the wheel opening. Completed ❑
23. Remove the paper backing from the pinstripe tape. Completed ❑
24. Position the pinstripe tape so that it is 1″ from the wheel well or wheel well arch at the lower front area. Completed ❑
25. Lightly apply the pinstripe tape around the wheel well, maintaining the 1″ distance. Completed ❑
26. Step back and look at the pinstripe tape. Reposition the pinstripe tape as needed to maintain a consistent space between the wheel well or arch and the tape. Completed ❑
27. Press down on the stripe to lock it into position. Completed ❑
28. Remove the clear film from the pinstripe tape. Completed ❑
29. Press the pinstripe tape into position to ensure proper adhesion. Completed ❑

Making a Stripe Point

30. Clean off a fender with soap and water. Then wipe it down with wax-and-grease remover. Completed ❑
31. Lightly apply a two-stripe pinstripe tape to the fender. Completed ❑
32. Remove the clear film from the pinstripe tape. Completed ❑
33. Pull one of the stripes from the vehicle and move it toward the other stripe. Completed ❑
34. Pull the second stripe from the vehicle and move it toward the first stripe so that a point is formed with the two stripes. Completed ❑
35. When the stripes are correctly positioned, press down on the stripes to lock them in place. Completed ❑

Copyright by Goodheart-Willcox Co., Inc. May not be reproduced or posted to a publicly accessible website.

Splice a Stripe Break

36. Clean off a panel with soap and water. Then wipe it down with wax-and-grease remover. Install a piece of 1/4″ or 1/8″ pinstripe tape.   Completed ❑

37. Cut off half of this stripe. This simulates a stripe break.   Completed ❑

38. Position a new stripe so that it overlaps the existing stripe by more than 1″.   Completed ❑

39. Place a razor blade on the new stripe at an angle, not vertically, and cut the new stripe at the 1″ mark.   Completed ❑

40. Press down on the new stripe to lock it in place.   Completed ❑

Laying Down a Paint Pinstripe

41. Clean the panel where the stripe will be made with soap and water. Then wipe it down with wax-and-grease remover.   Completed ❑

42. Measure and cut enough 1/8″ fine-line masking tape to cover the length of the panel. Apply the tape to the panel and press the tape in place. See **Figure 23-3**.   Completed ❑

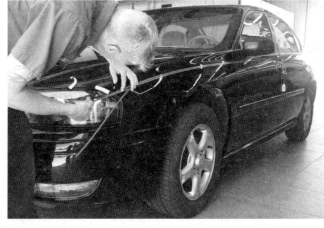

Figure 23-3. The replacement fender has been repainted and needs a new pinstripe to match the one on the door and quarter panel. A new pinstripe will be painted between the lines of fine-line masking tape the technician is applying.

43. Apply 1/4″ masking tape immediately adjacent to both sides of the 1/8″ masking tape. Apply 3/4″ masking tape to the 1/4″ masking tape.   Completed ❑

44. Remove the 1/8″ tape.   Completed ❑

45. Lightly scuff the area that was exposed when the 1/8″ tape was removed.   Completed ❑

46. Blow off the dust from the scuffed area.   Completed ❑

47. Clean off any remaining dust with a tack rag.   Completed ❑

48. Mix stripe paint.   Completed ❑

49. Use a fine-bristled paintbrush to apply the stripe paint to the taped area.   Completed ❑

50. Remove the masking tape *before* the stripe paint dries.   Completed ❑

Instructor's Initials _____

Date _____